Tools for High Performance Computing 2011

Holger Brunst • Matthias S. Müller
Wolfgang E. Nagel • Michael M. Resch
Editors

Tools for High Performance Computing 2011

Proceedings of the 5th International Workshop on Parallel Tools for High Performance Computing, September 2011, ZIH, Dresden

Springer

Editors
Holger Brunst
Matthias S. Müller
Wolfgang E. Nagel
Zentrum für Informationsdienste
und Hochleistungsrechnen (ZIH)
Technische Universität Dresden
01062 Dresden
Germany

Michael M. Resch
Höchstleistungsrechenzentrum
Stuttgart (HLRS)
Universität Stuttgart
Nobelstraße 19
70569 Stuttgart
Germany

Front cover figure: MPI communication pattern of a 3D cloud simulation on 512 CPU cores with dynamic load balancing. Hilbert space-filling curves are used for the distribution of the simulation data.

ISBN 978-3-642-31475-9 ISBN 978-3-642-31476-6 (eBook)
DOI 10.1007/978-3-642-31476-6
Springer Heidelberg New York Dordrecht London

Library of Congress Control Number: 2012948084

Mathematics Subject Classification (2010): 68-06, 68Q85, 68Q60, 68N19, 68U99, 94A99, 68N99

Printed on acid-free paper

Springer is part of Springer Science+Business Media (www.springer.com)

Preface

In the pursuit of maintaining exponential growth in the performance of high-performance computers, the HPC community is currently targeting Exascale systems. The initial planning for Exascale already started when the first Petaflop system was delivered. Many challenges need to be addressed to reach the required performance level. Scalability, energy efficiency, and fault-tolerance need to be increased by orders of magnitude. The goal can only be achieved when advanced hardware is combined with a suitable software stack. In fact, the importance of software is rapidly growing. As a result, many international projects focus on the necessary software. The International Exascale Software Project (IESP), the European Exascale Software Initiative (EESI), and the Virtual Institute for High Productivity Supercomputing (VI-HPS) are examples. They all share the view that tools are of components of the software stack.

The Parallel Tools Workshop that took place in Dresden on September 26–27, 2011, is the fifth in a series of workshops that started in 2007 at the High Performance Computing Center Stuttgart (HLRS). The goal of this series is to bring together tool developers and users from science and industry in an interactive environment. Participants from research and developers from science and industry were invited to this interactive workshop which attracted scientists from all over the world.

This year's presentations have been in the fields of System Management, Parallel Debugging and Performance Analysis from a wide range of scientific and industrial tool developers. This includes tools from vendors such as Allinea, ClusterVision, Intel, Rogue Wave Software, and SysFera, as well as research institutions, including Technische Universität Dresden, Universität Erlangen, University of Oregon, and Rice University. Contribution from research and computer centers came from Barcelona Supercomputing Center, Research Center Jülich, Karlsruhe Institute of Technology, Lawrence Berkeley National Laboratory, Lawrence Livermore National Laboratory, Pacific Northwest National Laboratory, and the High Performance Computing Center Stuttgart.

Dresden, Germany

Holger Brunst
Matthias S. Müller
Wolfgang E. Nagel
Michael M. Resch

Contents

Contributors

Hartwig Anzt Engineering Mathematics and Computing Lab (EMCL), Karlsruhe Institute of Technology (KIT), Karlsruhe, Germany

Werner Augustin Engineering Mathematics and Computing Lab (EMCL), Karlsruhe Institute of Technology (KIT), Karlsruhe, Germany

Martin Baumann Engineering Mathematics and Computing Lab (EMCL), Karlsruhe Institute of Technology (KIT), Karlsruhe, Germany

Abhinav Bhatele Center for Applied Scientific Computing, Lawrence Livermore National Laboratory, Livermore, CA, USA

Scott Biersdorff University of Oregon, Eugene, OR, USA

Steffen Brinkmann HLRS, Stuttgart, Germany

Peer-Timo Bremer Center for Applied Scientific Computing, Lawrence Livermore National Laboratory, Livermore, CA, USA

Kevin Coulomb SysFera, Lyon, France

Augustin Degomme INRIA Rhône-Alpes, Grenoble, France

Kai Diethelm Forschungszentrum Jülich, Jülich, Germany

Dominic Eschweiler Forschungszentrum Jülich, Jülich, Germany

Shiqing Fan High Performance Computing Center Stuttgart (HLRS), Stuttgart, Germany

Mathieu Faverge Innovative Computing Laboratory, University of Tennessee, Knoxville, TN, USA

Todd Gamblin Center for Applied Scientific Computing, Lawrence Livermore National Laboratory, Livermore, CA, USA

Markus Geimer Forschungszentrum Jülich, Jülich, Germany

Thomas Gengenbach Engineering Mathematics and Computing Lab (EMCL), Karlsruhe Institute of Technology (KIT), Karlsruhe, Germany

Michael Gerndt Technische Universität München, München, Germany

Judit Giménez Barcelona Supercomputing Center, Universitat Politècnica de Catalunya, Barcelona, Catalunya, Spain

Chris Gottbrath Rogue Wave Software, Boulder, CO, USA

José Gracia HLRS, Stuttgart, Germany

Georg Hager Erlangen Regional Computing Center (RRZE), Friedrich-Alexander Universität Erlangen-Nürnberg, Erlangen, Germany

Tobias Hahn Engineering Mathematics and Computing Lab (EMCL), Karlsruhe Institute of Technology (KIT), Karlsruhe, Germany

Andreas Helfrich-Schkarbanenko Engineering Mathematics and Computing Lab (EMCL), Karlsruhe Institute of Technology (KIT), Karlsruhe, Germany

Vincent Heuveline Engineering Mathematics and Computing Lab (EMCL), Karlsruhe Institute of Technology (KIT), Karlsruhe, Germany

Kevin Huck Barcelona Supercomputing Center, Universitat Politècnica de Catalunya, Barcelona, Catalunya, Spain

Katherine Isaacs Center for Applied Scientific Computing, Lawrence Livermore National Laboratory, Livermore, CA, USA

Guido Juckeland Center for Information Services and High Performance Computing (ZIH), Technische Universität Dresden, Dresden, Germany

Rainer Keller High Performance Computing Center Stuttgart (HLRS), Stuttgart, Germany

HLRS, Stuttgart, Germany

Eva Ketelaer Engineering Mathematics and Computing Lab (EMCL), Karlsruhe Institute of Technology (KIT), Karlsruhe, Germany

Andreas Knüpfer ZIH, TU Dresden, Dresden, Germany

Jesús Labarta Barcelona Supercomputing Center, Universitat Politècnica de Catalunya, Barcelona, Catalunya, Spain

Chee Wai Lee University of Oregon, Eugene, OR, USA

Joshua A. Levine Scientific Computing and Imaging Institute University of Utah, Salt Lake City, UT, USA

Germán Llort Barcelona Supercomputing Center, Universitat Politècnica de Catalunya, Barcelona, Catalunya, Spain

Daniel Lorenz Technische Universität München, München, Germany

Royd Lüdtke Rogue Wave Software, Boulder, CO, USA

Dimitar Lukarski Engineering Mathematics and Computing Lab (EMCL), Karlsruhe Institute of Technology (KIT), Karlsruhe, Germany

Allen Malony University of Oregon, Eugene, OR, USA

John Mellor-Crummey Department of Computer Science, Rice University, Houston, TX, USA

Dieter an Mey RWTH Aachen, Aachen, Germany

Wolfgang E. Nagel ZIH, TU Dresden, Dresden, Germany

Andrea Nestler Engineering Mathematics and Computing Lab (EMCL), Karlsruhe Institute of Technology (KIT), Karlsruhe, Germany

Christoph Niethammer HLRS, Stuttgart, Germany

Yury Oleynik Technische Universität München, München, Germany

Valerio Pascucci Scientific Computing and Imaging Institute University of Utah, Salt Lake City, UT, USA

Peter Philippen Forschungszentrum Jülich, Jülich, Germany

Michael Resch High Performance Computing Center Stuttgart (HLRS), Stuttgart, Germany

Sebastian Ritterbusch Engineering Mathematics and Computing Lab (EMCL), Karlsruhe Institute of Technology (KIT), Karlsruhe, Germany

Staffan Ronnas Engineering Mathematics and Computing Lab (EMCL), Karlsruhe Institute of Technology (KIT), Karlsruhe, Germany

Christian Rössel Forschungszentrum Jülich, Jülich, Germany

Pavel Saviankou Forschungszentrum Jülich, Jülich, Germany

Michael Schick Engineering Mathematics and Computing Lab (EMCL), Karlsruhe Institute of Technology (KIT), Karlsruhe, Germany

Dirk Schmidl RWTH Aachen, Aachen, Germany

Mareike Schmidtobreick Engineering Mathematics and Computing Lab (EMCL), Karlsruhe Institute of Technology (KIT), Karlsruhe, Germany

Martin Schulz, Center for Applied Scientific Computing, Lawrence Livermore National Laboratory, Livermore, CA, USA

Harald Servat Barcelona Supercomputing Center, Universitat Politècnica de Catalunya, Barcelona, Catalunya, Spain

Sameer Shende University of Oregon, Eugene, OR, USA

Wyatt Spear University of Oregon, Eugene, OR, USA

Chandramowli Subramanian Engineering Mathematics and Computing Lab (EMCL), Karlsruhe Institute of Technology (KIT), Karlsruhe, Germany

Nathan R. Tallent Pacific Northwest National Laboratory, Richland, WA, USA

François Trahay Institut Télécom, Télécom SudParis, Evry Cedex, France

Ronny Tschüter ZIH, TU Dresden, Dresden, Germany

Jan Treibig Erlangen Regional Computing Center (RRZE), Friedrich-Alexander Universität Erlangen-Nürnberg, Erlangen, Germany

Michael Wagner ZIH, TU Dresden, Dresden, Germany

Jan-Philipp Weiss Engineering Mathematics and Computing Lab (EMCL), Karlsruhe Institute of Technology (KIT), Karlsruhe, Germany

Gerhard Wellein Erlangen Regional Computing Center (RRZE), Friedrich-Alexander Universität Erlangen-Nürnberg, Erlangen, Germany

Bert Wesarg ZIH, TU Dresden, Dresden, Germany

Florian Wilhelm Engineering Mathematics and Computing Lab (EMCL), Karlsruhe Institute of Technology (KIT), Karlsruhe, Germany

Martin Wlotzka Engineering Mathematics and Computing Lab (EMCL), Karlsruhe Institute of Technology (KIT), Karlsruhe, Germany

Felix Wolf German Research School for Simulation Sciences, Aachen, Germany

Chapter 1
Creating a Tool Set for Optimizing Topology-Aware Node Mappings

Martin Schulz, Abhinav Bhatele, Peer-Timo Bremer, Todd Gamblin, Katherine Isaacs, Joshua A. Levine, and Valerio Pascucci

Abstract Modern HPC systems, such as Cray's XE and IBM's Blue Gene line, feature sophisticated network architectures, often in the form of high dimensional tori. In order to fully exploit the performance of these systems, it is necessary to carefully map an application's communication structure to the underlying network topology. In this step, both latency (i.e., physical distance between nodes) and bandwidth (i.e., number of concurrently used links) have to be taken into account, leading to mappings that are often non-intuitive. To help developers with this complex problem, we are developing a set of tools that aim at helping users understand the communication behavior of their codes, map them onto the network architecture, and create better-performing topology-aware node mappings. In this paper, we present initial steps towards this goal, including a measurement environment capturing both communication patterns and network metrics within the same run, a methodology to compare these measurements, and a visualization tool that helps users understand the impact of their application's characteristics on the network behavior.

1.1 Motivation and Background

Modern HPC architectures typically feature high dimensional tori as their interconnection network. Some examples are the IBM Blue Gene line, Cray's XT/XE systems, and Fujitsu's K computer. Because of their constant per-node link count,

M. Schulz (✉) · A. Bhatele · P.-T. Bremer · T. Gamblin · K. Isaacs
Center for Applied Scientific Computing, Lawrence Livermore National Laboratory, Livermore, CA 94550, USA
e-mail: schulzm@llnl.gov; bhatele@llnl.gov; bremer5@llnl.gov; tgamblin@llnl.gov; isaacs6@llnl.gov

J.A. Levine · V. Pascucci
Scientific Computing and Imaging Institute University of Utah, Salt Lake City, UT 84112, USA
e-mail: jlevine@sci.utah.edu; pascucci@sci.utah.edu

H. Brunst et al. (eds.), *Tools for High Performance Computing 2011*,
DOI 10.1007/978-3-642-31476-6_1, © Springer-Verlag Berlin Heidelberg 2012

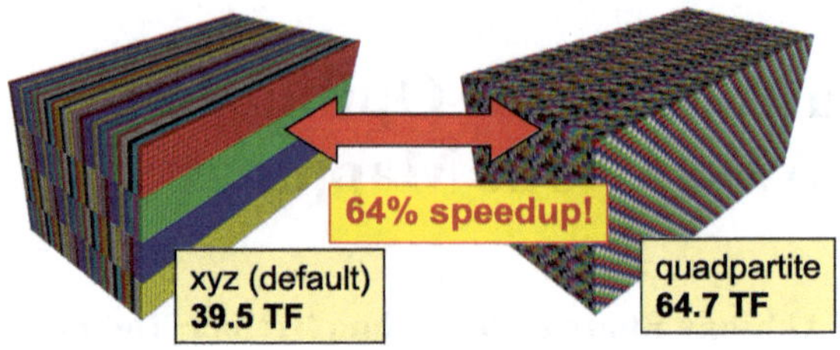

Fig. 1.1 Application performance of an FPMD application on 64 K nodes of Blue Gene/L using the default (*left*) and the optimized mapping (*right*). Blocks of the same color denote nodes assigned to the same row of the core data structure, a dense 2D matrix

these networks are relatively easy to deploy and scale. However, the resulting high network diameter creates performance and scaling problems for applications. Thus, it is critical to lay out the tasks of an application on the underlying network topology carefully, so that communication is efficient for the specific patterns the application exhibits.

The topology mapping problem is well-documented in the literature [2–5, 10, 18]. As one example, during the optimization process of a first principle molecular dynamics (FPMD) application [10], we found that the performance difference between the default and the best performing layout was 64 %. Further, the best mapping was contrary to prior expectations and non-intuitive (see Figure 1.1).

In order to enable such optimizations, we need tools that help users understand the communication patterns of their application, how they map onto the hardware topology, how this impacts performance, and ultimately provide efficient node mappings for their applications. To achieve this goal, we make use of performance data gathered in multiple independent domains to provide the user with informative perspectives on the performance of their code [15]. Further, we can map data gathered in one domain to another domain enabling us to compare and correlate the performance data and features found in different domains. In particular, we focus on data from the communication domain, representing the application's communication pattern, and the hardware domain, in which we gather network statistics. This allows us to understand the impact of an application's communication behavior on the network for a particular layout.

We present initial steps towards a novel tool set to explore and optimize topology-aware node mappings for HPC applications, using information mapped and correlated between multiple performance domains. We make the following contributions:

- We show the use of a three-domain model of performance data for studying node mappings in HPC systems.
- We present a flexible measurement tool set that can be used to concurrently track data from multiple measurement domains.
- We introduce a technique to correlate network measurements with application communication behavior.

- We show preliminary results of an Algebraic Multigrid (AMG) solver.

The remainder of this paper is organized as follows: Sect. 1.2 introduces the underlying domain model as well as the necessary data mappings. Section 1.3 discusses how we gather and interpret the data, Sect. 1.4 shows first preliminary results, Sect. 1.5 briefly discusses related work, and Sect. 1.6 contains concluding remarks and an overview of future work.

1.2 Gaining Insight from Multiple Perspectives

Performance data can be gathered and analyzed in multiple domains. In previous work [15], we provided a structured approach for the three most relevant performance data domains. In this section, we briefly introduce this model and show how it applies to the node mapping problem.

1.2.1 The HAC *Model*

Figure 1.2 shows the basic concept behind our domain model, the *HAC* model. We identify the three most relevant domains for performance analysis, namely the application, the hardware, and the communication domain. The application domain is used by the application scientist to express the actual problem being computed. For scientific simulations it often represents some physical simulation space on which the problem is defined, although more abstract representations are also possible (e.g., for graph algorithms or sparse matrix representations in solver libraries). The hardware domain expresses the structure of the underlying architecture, e.g., a torus with multicore nodes and is often used for hardware related measurements, such as hardware counters. The communication domain is used to express the code's communication pattern, often represented as a graph with nodes representing application tasks (e.g., MPI processes) and edges showing communication between nodes.

For each of the three domains we can use specialized analysis techniques that allow us to extract features from the respective data sets. Additionally, we can define mappings for data between domains allowing us to compare data from multiple domains for further analysis. For example, we can map performance data gathered in the hardware domain, such as hardware counter data, into the application domain and display it concurrently with simulated features from the application. In our previous work [15] we have demonstrated that this can show correlations between simulated features and measured performance data allowing us a clear attribution of performance data and establishing clear baselines.

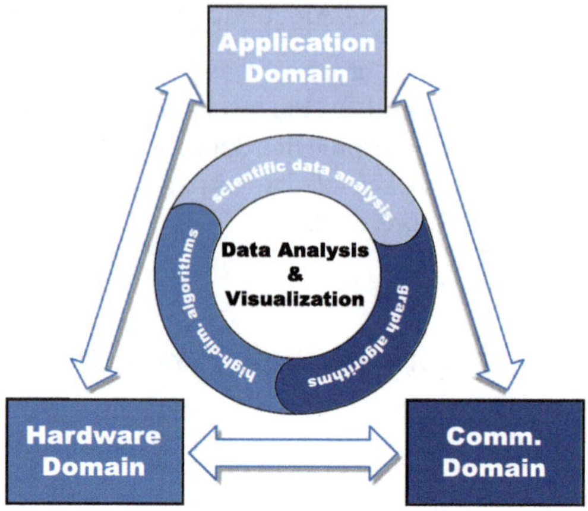

Fig. 1.2 Interplay between the three critical domains for performance analysis

1.2.2 Domains Relevant for Node Mappings

For the node mapping problem, we focus on two of the three domains: the communication domain showing an application's communication behavior and the hardware domain showing its network behavior. By contrasting the data from these two domains we can analyze the impact of the application's characteristics on the network behavior and gain the necessary insight to optimize performance.

In the communication domain we are interested in all communication executed by the application, in our case in the form of MPI messages. Conceptually, we gather the full communication graph, although in reality we restrict ourselves to a statistical summary of the data which is sufficient for our purposes. In the hardware domain we must acquire the data showing the actual communication executed in the system. For this work we focus first on Blue Gene systems, which provide a set of performance counters to monitor the actual number of packets exchanged on each hardware link in the system.

1.2.3 Mapping Communication to Hardware Data

We must establish a mapping between the two domains in order to allow a direct comparison of hardware and communication data. This mapping has two components: we first map MPI processes to physical nodes in the hardware domain. This is the actual node mapping we are aiming to optimize and hence we will evaluate how changing this mapping affects performance.

Second, for a concrete node mapping, we map the performance data from one domain to the other. Ideally, we would map the hardware domain data to the communication domain, since this is the most relevant for the user. Unfortunately, though, this is impossible because each hardware link is potentially used by several MPI communication pairs, since links are shared for all packets passing through the connected nodes. Therefore, we cannot cleanly correlate the hardware data back to individual MPI message events.

We therefore reverse the mapping and map the MPI communication behavior to the network architecture. We do this by mapping each message individually, assuming no contention, interference, or bandwidth limits, and with that we emulate the hardware behavior under ideal load. As a result, we gain a best case view of how the application's communication under the given node mapping behaves on the system. We can contrast this best case estimate with the actual measurements showing where behavior deviates from the best case estimate and hence where bottlenecks manifest themselves during execution.

1.3 Creating a Flexible Measurement Environment

Our approach requires the ability to gather data from multiple domains concurrently, since only this keeps the data comparable and avoids cross-execution variations. We implement this on top of the P^NMPI infrastructure, and develop a series of plugin modules for the actual data acquisition. We further define structured, self-explanatory data formats to store the data for post-processing in our analysis and visualization tools.

1.3.1 Concurrent Measurements Using P^NMPI

As most tools targeting MPI, we rely on the MPI Profiling Interface (PMPI), which allows tools to transparently intercept invocations to MPI routines and with that to establish wrappers around MPI calls to gather execution information. However, the usage of this interface is limited to a single tool. We therefore rely on the P^NMPI infrastructure [14], which virtualizes this interface and enables the concurrent execution of an arbitrary number of PMPI tools, in the extreme case even without changes to the tool or the application. Figure 1.3 illustrates the principle behind P^NMPI.

In the following we refer to PMPI tools used within P^NMPI as tool modules, since they are used as plugin modules in the overall infrastructure. The modules used for a specific run are specified in an ASCII configuration file, which P^NMPI loads at startup and then uses to locate, load, and instantiate its modules.

For our experiments we implement the data acquisition in the two domains as separate P^NMPI modules and load them dynamically at runtime. Both are executed independently and can gather data in their respective domains.

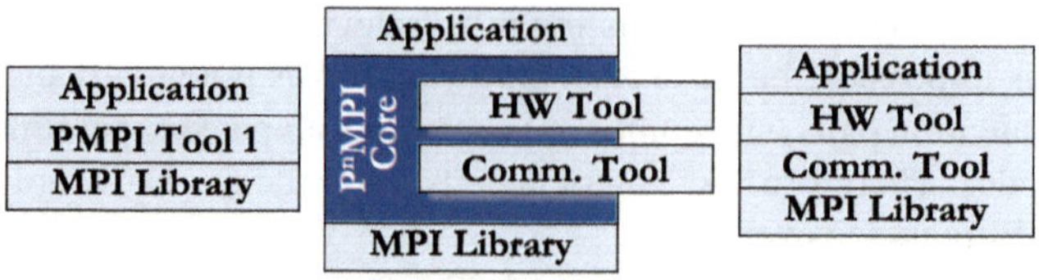

Fig. 1.3 Usage of the MPI profiling interface (*left*); combining two measurement tools using P^NMPI (*middle*); effective tool stack as seen by the application (*right*)

1.3.2 Gathering Data in the Hardware Domain

In the hardware domain we collect performance counter data from each node in the application's partition. In particular, we gather the number of packets sent in each direction of the torus from each node along with some non-network metrics such as cache misses or floating point operations. We implement this module on top of IBM's Universal Performance Counter (UPC) interface. At program termination, we gather the data to rank zero and store the result as a table with a row for each rank and a column for each hardware counter metric.

1.3.3 Gathering Data in the Communication Domain

The data in the communication domain represents the communication patterns of the application. We gather this information in the form of a communication matrix that stores aggregated information for all MPI rank pairs that exchange messages during the application's execution. In particular, we store the number of messages sent between two ranks as well as the total number of bytes. Further, we distinguish the data for three different messages sizes to capture data for different communication protocols on the Blue Gene systems. The latter is necessary, since different communication protocols lead to different network traffic and hence emulating this traffic requires separate handling for each protocol.

1.3.4 Phase Attribution

For both domains we further need to refine the data gathered by the two modules to expose phase behavior. For this we instrument the application with `MPI_Pcontrol` calls to specify application specific phase boundaries. We use the `level` argument to `MPI_Pcontrol` to indicate the ID of new phases and extend both modules to recognize these phase transitions and to annotate any performance data with the ID of the phase during which it was collected.

```
---
- !!python/tuple [mpirank, int32]
- !!python/tuple [phase, int32]
- !!python/tuple [x, int32]
- !!python/tuple [y, int32]
...
0 1 4 5
1 1 8 9
0 2 1 1
1 2 6 3
```

Fig. 1.4 Example of a YAML file (header and body) showing data for a 2D coordinates (x/y) with data for different ranks (0, 1) and application phases (with IDs 1, 2)

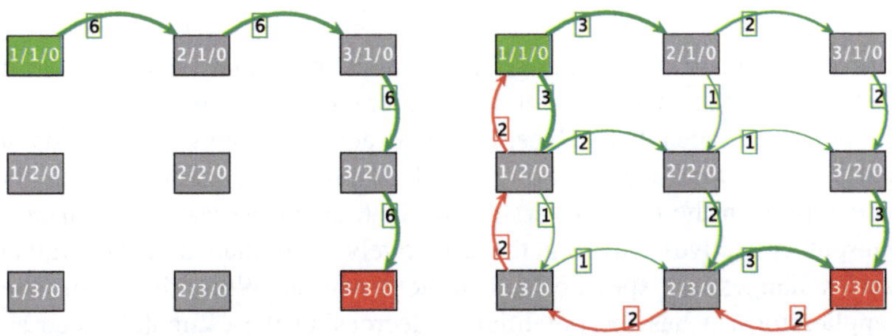

Fig. 1.5 Message traffic on a part of a plane of BG/P nodes for a short message (*left*) and a long message (*right*). *Green arrows* indicate packets sent in the positive direction of the torus network, *red arrows* packets in the negative

1.3.5 Data Storage in Structured YAML Files

The results from both modules are stored in the form of a structured file using ASCII together with a YAML header [7] describing the contents. An example is shown in Fig. 1.4. This enables our tools to interpret the data, opens our tools to multiple data sources, and allows us to implicitly document the data represented in the file.

1.3.6 Investigating Best Case Mappings

Once gathered, we need to map the data from one of the domains to another. As mentioned in Sect. 1.2.3, we do this by mapping the effect of all messages observed in the communication domain individually onto the hardware domain. To enable this, we first studied the network traffic for individual MPI messages (see Fig. 1.5) and used this data to construct an accurate model for individual messages. To capture the correct behavior we distinguish short and long messages and model them separately.

We then apply this model to every communication pair observed in the communication domain and recorded in the communication matrix during the execution of our target application and add up the resulting number of modeled packets. This gives us an overall estimate of the total traffic in the system. While this estimate is not accurate since it does not take contention and overlapping and interfering traffic into account, it still gives a best case estimate of traffic as if the traffic was executed on a system with unbounded communication capabilities. Any difference between this estimate and the actual observed traffic is therefore caused by system limitations or bottlenecks and therefore gives an indication of problems with the underlying node mappings.

1.3.7 Approximating Collectives

The current approach only covers point to point messages, since we can isolate their behavior on the network. Collective operations are implemented as a set of point to point messages themselves, but the exact algorithm used depends on both the underlying network topology—and with that the node mapping—and the MPI implementation. In the current version of our tool set we use an idealized model to represent collectives showing the effectively communicated data rather than the data exchanges in a specific MPI implementation. While this is sufficient for most applications, it has the potential of decreasing the estimate's accuracy. We are currently working on extending the approach using probabilistic models for collectives to avoid this drawback.

1.4 Preliminary Results

To demonstrate our techniques, we apply them to an Algebraic Multigrid (AMG) solver from the hypre library [8]. This codes solves a system of sparse linear equations using an iterative approach and operates on a V cycle, which means it starts with a fine grid, coarsens it through multiple levels until it can be solved directly, and interpolates the coarse grain solution back to the fine grain grid. This process is then repeated until the system converges to a stable solution.

While finer levels of the AMG computation are compute-dominated and have regular communication structures with a limited set of neighbors, coarser levels are communication bound and exhibit more random and input-dependent communication pattern with a large set of neighbors [9].

We execute the AMG code on 512 nodes of a Blue Gene/P machine at LLNL, which are organized as a $8 \times 8 \times 8$ torus. We extract both communication and hardware counter data using our measurement infrastructure. The top row of graphs in Fig. 1.6 shows the measured link activity visualized on top of the underlying physical 3D torus structure for two node mappings and for specific AMG levels; the middle row of graphs shows the communication data mapped into the hardware domain using our best case network emulation approach introduced above, and

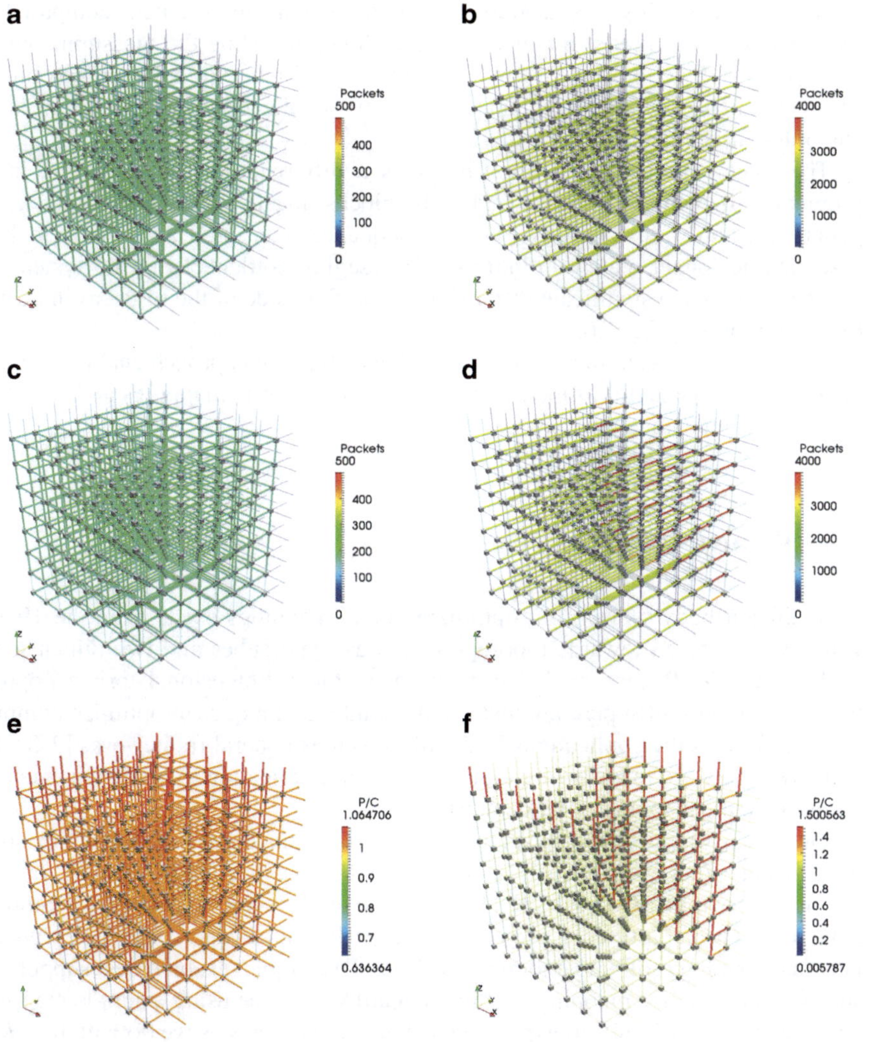

Fig. 1.6 Measured vs. emulated network traffic for two node mappings of AMG on 512 nodes of BG/P. (**a**) 8 × 8 × 8, level 2, counter data. (**b**) 2 × 4 × 16, level 1, counter data. (**c**) 8 × 8 × 8, level 2, emulated data. (**d**) 2 × 4 × 16, level 1, emulated data. (**e**) 8 × 8 × 8, level 2, emulated/counter. (**f**) 2 × 4 × 16, level 1, emulated/counter

the bottom row shows the ratio between the emulated data and the hardware measurements.

The left column of the data shows results from a run in which the data set is decomposed into an 8 × 8 × 8 problem and hence matches the underlying topology. We can clearly see that the measured and estimated data are close to each other, indicating a mostly bottleneck free execution. However, looking at the ratio graph, we can see a roughly 7 % degradation of the actual measurement compared to the ideal best case model for links in the Z direction, while the measurements in the X and Y direction don't show a significant difference (red vs. orange links). This asymmetry is unexpected since neither the application nor the system favor any of the three dimensions.

The right column of the Fig. 1.6 shows a different picture. Here, the input problem is decomposed into $2 \times 4 \times 16$ blocks and hence does not match the underlying hardware topology. As a consequence, measured and emulated best case data no longer match up indicating messaging bottlenecks in the system. In particular, we see a significant difference on the right side of the torus, as indicated by the red links in Fig. 1.6f.

Overall, our measurement, analysis, and visualization approach enables users to understand the behavior of their code under different node mappings and based on that optimize the node mappings.

1.5 Related Work

A significant body of work on optimizing node mappings exists from the 1980s, which was targeted towards topologies such as hypercubes and two-dimensional grids [1, 6, 12]. Recent work has focused on three-dimensional tori, a popular topology for current supercomputers, and includes techniques to optimize communication for specific applications [4, 5, 10] as well as general frameworks [2, 3, 18]. However, there are few existing tools to map communication to hardware data and to visualize the mapping of tasks on processors.

Individual application developers have developed or exploited existing visualization frameworks such as ParaView to visualize task mapping on the physical network. SCALASCA [17] and its predecessor Kojak [13] feature a data browser that presents projections of performance data onto the physical network topology of supercomputers. The TAU tool set includes support to specify and visualize mappings within the hardware space using a simple mapping language [16], but it fails to exploit inter-domain mappings as we do with the *HAC* model. The Projections visualization framework [11] has capabilities to show the average number of hops/links that messages originating from a processor travel on a 3D torus/mesh. However, it does not have 3D views to visualize this data.

1.6 Conclusions and Future Work

Optimizing the mapping of application processes to nodes in the physical topology of a machine is a critical performance optimization step on today's HPC systems. Without it, applications might pay high messaging costs both in terms of latency and bandwidth. In this work, we presented an initial step towards a general tool set for

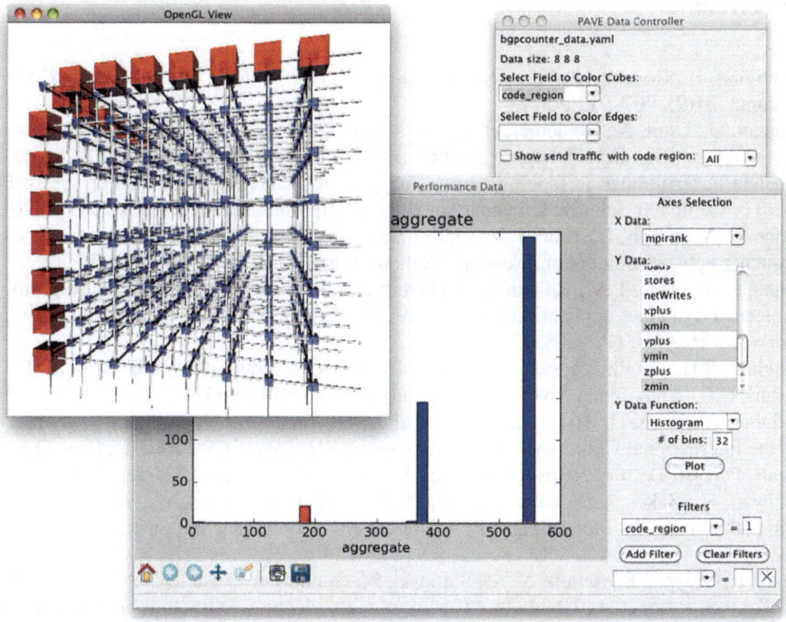

Fig. 1.7 Screenshot of the Boxfish tool showing the data control panel (*right*), statistical data on link usage (*middle*), and the 3D torus of Blue Gene/P with the nodes highlighted that correspond to the selected statistical data (*left*)

optimizing these mappings. Our approach enables users to correlate communication data with actual observations on the system leading to the detection of bottlenecks in the communication subsystem. We introduced the base measurement infrastructure and explored the overall approach using an Algebraic Multigrid solver as a case study.

We are currently in the process of extending this work in several directions. We are working on additional application studies using a 3D laser-plasma interaction code at LLNL as well as an adaptive mesh refinement library. In both cases, the communication, and consequently the node mappings, have been identified as one of the critical bottlenecks. We will study these codes both in a static scenario, where the partition size and shape is known a priori, and in a dynamic scenario, where the scheduling system decides on the partition shape at job startup. Further, we are currently developing a flexible data analysis and visualization tool, Boxfish (see Fig. 1.7), that will enable users to interactively analyze their codes' behavior and will guide mapping optimizations.

Acknowledgements This work was partially performed under the auspices of the U.S. Department of Energy by Lawrence Livermore National Laboratory under Contract DE-AC52-07NA27344 (LLNL-CONF-508551).

References

1. Aleliunas, R., Rosenberg, A.L.: On embedding rectangular grids in square grids. IEEE Trans. Comput. **31**(9), 907–913 (1982)
2. Bhanot, G., Gara, A., Heidelberger, P., Lawless, E., Sexton, J.C., Walkup, R.: Optimizing task layout on the Blue Gene/L supercomputer. IBM J. Res. Dev. **49**(2/3), 489–500 (2005)
3. Bhatele, A.: Automating topology aware mapping for supercomputers. Ph.D. thesis, Department of Computer Science, University of Illinois. http://hdl.handle.net/2142/16578 (2010)
4. Bhatele, A., Bohm, E., Kale, L.V.: Optimizing communication for charm++ applications by reducing network contention. Concurr. Comput. Pract. Exp. **23**(2), 211–222 (2011)
5. Bhatele, A., Kalé, L.V., Kumar, S.: Dynamic topology aware load balancing algorithms for molecular dynamics applications. In: 23rd ACM International Conference on Supercomputing, Yorktown Heights, NY (2009)
6. Bokhari, S.H.: On the mapping problem. IEEE Trans. Comput. **30**(3), 207–214 (1981)
7. Evans, C.C.: The official YAML web site. http://yaml.org/ (2011)
8. Falgout, R., Yang, U.: Hypre: a library of high performance preconditioners. In: Proceedings of the International Conference on Computational Science (ICCS), Amesterdam, The Netherlands. Part III, Lecture Notes in Computer Science, vol. 2331, pp. 632–641 (2002)
9. Gahvari, H., Baker, A., Schulz, M., Yang, U.M., Jordan, K., Gropp, W.: Scalable fine-grained call path tracing. In: Proceedings of the International Conference on Supercomputing, Tucson, AZ (2011)
10. Gygi, F., Draeger, E., Schulz, M., de Supinski, B., Gunnels, J., Austel, V., Sexton, J., Franchetti, F., Kral, S., Lorenz, J., Überhuber, C.: Large-scale electronic structure calculations of high-Z metals on the BlueGene/L platform. In: Proceedings of IEEE/ACM Supercomputing 2006 (SC06), Tamp, FL (2006)
11. Kale, L.V., Zheng, G., Lee, C.W., Kumar, S.: Scaling applications to massively parallel machines using projections performance analysis tool. Future Gener. Comput. Syst. **22**, 347–358 (2006). Special issue on: Large-scale system performance modeling and analysis
12. Lee, S.-Y., Aggarwal, J.K.: A mapping strategy for parallel processing. IEEE Trans. Comput. **36**(4), 433–442 (1987)
13. Mohr, B., Wolf, F.: KOJAK – a tool set for automatic performance analysis of parallel programs. In: Proceedings of the International Conference on Parallel and Distributed Computing (Euro-Par 2003), Klagenfurt, Austria, pp. 1301–1304 (2003)
14. Schulz, M., de Supinski, B.R.: $P^N MPI$ Tools: a whole lot greater than the sum of their parts. In: Proceedings of Supercomputing 2007 (SC07), Reno, NV (2007)
15. Schulz, M., Levine, J.A., Bremer, P.T., Gamblin, T., Pascucci, V.: Interpreting performance data across intuitive domains. In: International Conference on Parallel Processing (ICPP), Taipei City, Taiwan (2011)
16. Spear, W., Malony, A.D., Lee, C.W., Biersdorff, S., Shende, S.: An approach to creating performance visualizations in a parallel profile analysis tool. In: Workshop on Productivity and Performance (PROPER), Bordeaux, France (2011)
17. Wolf, F., Wylie, B., Abraham, E., Becker, D., Frings, W., Fuerlinger, K., Geimer, M., Hermanns, M.A., Mohr, B., Moore, S., Szebenyi, Z.: Usage of the SCALASCA toolset for scalable performance analysis of large-scale parallel applications. In: Proceedings of the 2nd HLRS Parallel Tools Workshop, Stuttgart. http://www.cs.utk.edu/~karl/research/pubs/WOLF_2008_Scalasca.pdf (2008)
18. Yu, H., Chung, I.H., Moreira, J.: Topology mapping for blue Gene/L supercomputer. In: SC '06: Proceedings of the 2006 ACM/IEEE Conference on Supercomputing, p. 116. ACM, New York (2006). http://doi.acm.org/10.1145/1188455.1188576

Chapter 2
Using Sampling to Understand Parallel Program Performance

Nathan R. Tallent and John Mellor-Crummey

Abstract Developing scalable parallel applications for extreme-scale systems is challenging. The challenge of developing scalable parallel applications is only partially addressed by existing languages, compilers, and autotuners. As a result, manual performance tuning is often necessary to obtain high application performance. Rice University's HPCTOOLKIT is a suite of performance tools that supports innovative techniques for pinpointing and quantifying performance bottlenecks in fully optimized parallel programs with a measurement overhead of only a few percent. Many of these techniques were designed to leverage sampling for performance measurement, attribution, analysis, and presentation. This paper surveys some of HPCTOOLKIT's most interesting techniques and argues that sampling-based performance analysis is surprisingly versatile and effective.

Keywords Performance tools • Sampling • Call path profiling • Call path tracing • HPCTOOLKIT

2.1 Introduction

Because of the complexities of modern microprocessor-based supercomputers—e.g., scale, hierarchical parallelism, heterogeneity, deep memory hierarchies—typical scientific applications achieve only a small fraction of a supercomputer's peak performance. As a result, manual performance analysis and tuning is often

N.R. Tallent (✉)
Pacific Northwest National Laboratory, Richland, WA 99352, USA
e-mail: tallent@pnnl.gov

J. Mellor-Crummey
Department of Computer Science, Rice University, Houston, TX 77005, USA
e-mail: johnmc@rice.edu

H. Brunst et al. (eds.), *Tools for High Performance Computing 2011*,
DOI 10.1007/978-3-642-31476-6_2,

necessary to identify and resolve performance bottlenecks to avoid wasting precious computational resources.

Rice University's HPCToolkit [1, 19] is a suite of performance tools that has pioneered techniques for pinpointing and quantifying performance bottlenecks in fully optimized parallel programs with a measurement overhead of only a few percent [2, 5, 9, 13, 15, 22–24, 26–29]. Many of these techniques were designed to leverage sampling for performance measurement, attribution, analysis, and presentation. This paper surveys some of HPCToolkit's most interesting techniques and argues that sampling-based performance analysis is surprisingly versatile and effective.

Prima facie, sampling is both an obvious and questionable technique for performance analysis. Consider the challenge of collecting both accurate and detailed performance measurements. There are two basic techniques for collecting performance measurements: asynchronous sampling and synchronous instrumentation. Although there is an inverse relationship between accuracy and precision, the elasticity of that relationship varies with measurement technique. With instrumentation, collecting context-sensitive measurements at the level of a procedure can easily lead to hundreds or even thousands of percent overhead [3, 9, 23]; measuring at finer resolutions such as the loop, statement, or machine-instruction level is often infeasible. Efforts to address these problems include statically [4, 10, 21] or dynamically [12, 16, 21] omitting instrumentation, which leads to blind spots; adjusting measurements based on a model of how instrumentation affects execution [14]; and other specialized techniques [3, 11, 30]. We argue that, when possible, it is better to simply avoid instrumentation overhead and its associated distortion. To provide a more elastic trade-off between accuracy and precision than instrumentation can provide, HPCToolkit collects statistically representative measurements by periodically interrupting an execution using an asynchronous event trigger. Just as pollsters query a thousand voters to discern trends in a population of 300 million, asynchronously sampling the activity in a program execution can provide representative information about the program's behavior.

But if sampling appears an obvious choice, there are a number of considerations that make it questionable. First, the fact that asynchronous sampling interrupts an execution at arbitrary points poses certain difficulties. One difficulty is how to attribute measurements to source-level loops. With source code instrumentation such attribution is trivial (though costly). With sampling it is difficult because loops in object code are implemented using unstructured control flow and have been transformed by optimizing compilers. Another difficulty is collecting calling contexts using stack unwinding. Experience shows that unwinds commonly fail when using libunwind [17] or GLIBC's `backtrace()` routine [7] from within highly optimized code. Second, the largest supercomputers present challenges of their own. To avoid idleness in bulk synchronous parallel applications, supercomputer operating systems frequently attempt to minimize OS 'jitter' [18]. Unfortunately from the perspective of sampling, a standard way to minimize jitter is to forbid frequent software interrupts. Supercomputer kernels may also leave unimplemented the OS support needed to drive asynchronous sampling triggers. Third, it is often not clear how to use sampling to accomplish a primary goal of performance analysis: pinpointing root causes of bottlenecks. For instance, because sampling may miss the first (or any other) link in a chain of events, it may have limited diagnostic powers. More concretely, consider the problem of pinpointing the causes of lock contention. A straightforward solution uses instrumentation to time lock acquisitions and releases and attributes those measurements to their calling context. However, adding such instrumentation will lengthen critical sections and often exacerbate lock contention. Is it possible to use a sampling-based technique that can determine, for only a few percent of overhead, when a thread is working with a lock, working without a lock, or waiting for a lock? More to the point, is it possible to not just identify threads that wait for locks, but the threads and critical sections that *cause* waiting?

This paper is organized as follows. Section 2.2 summarizes HPCTOOLKIT's call path profiling capabilities. Section 2.3 explains how to pinpoint scaling bottlenecks using differential analysis. Section 2.4 presents techniques for pinpointing the causes (not effects) of bottlenecks. Section 2.5 describes how to use sampling for collection and presentation of call path traces. Section 2.6 summarizes HPC-TOOLKIT's data-centric attribution capabilities. Finally, Sect. 2.7 summarizes our conclusions. Because of space constraints and a retrospective perspective, we delegate a thorough discussion of related work to individual research papers.

2.2 Call Path Profiling

Experience shows that comprehensive performance analysis of modern modular software requires information about the full *calling context* in which costs are incurred. For instance, the costs incurred for calls to communication primitives (e.g., `MPI_Wait`) or code that results from instantiating C++ templates for data structures can vary widely depending upon their calling context. Consequently, HPCTOOLKIT uses asynchronous sampling to collect *call path profiles* that attribute costs to the full calling contexts in which they are incurred.

Collecting a call path profile requires capturing the calling context for each sample event. To capture the calling context for a sample event, HPCTOOLKIT's measurement tool `hpcrun` must be able to unwind the call stack at *any* point in a program's execution. Obtaining the return address for a procedure frame that does not use a frame pointer is challenging since the frame may dynamically grow (space is reserved for the caller's registers and local variables; the frame is extended with calls to `alloca`; arguments to called procedures are pushed) and shrink (space for the aforementioned purposes is deallocated) as the procedure executes. To cope with this situation, we developed a fast, on-the-fly binary analyzer that examines a routine's machine instructions and computes how to unwind a stack frame for the procedure [23]. For each address in the routine, there must be a recipe for how to unwind. Different recipes may be needed for different intervals of addresses within the routine. Each interval ends with an instruction that changes the state of the routine's stack frame. Each recipe describes (1) where to find the current frame's return address, (2) how to recover the value of the stack pointer for the caller's frame, and (3) how to recover the value that the base pointer register had in the caller's frame. Once we compute unwind recipes for a routine, we memoize them for later reuse.

To apply our binary analysis to compute unwind recipes, we must know where each routine starts and ends. When working with applications, one often encounters partially stripped libraries or executables that are missing information about function boundaries. To address this problem, we developed a binary analyzer that infers routine boundaries by noting instructions that are reached by call instructions or instructions following unconditional control transfers (jumps and returns) that are not reachable by conditional control flow.

HPCTOOLKIT's use of binary analysis for call stack unwinding has proven to be very effective, even for fully optimized code. A detailed study of our unwinder on versions of the SPEC CPU2006 benchmarks optimized with several different compilers showed that the unwinder was able to recover the calling context for all but a vanishingly small number of cases [23].

One particularly useful and novel feature of HPCTOOLKIT's call path profiles is that they are enriched with static source code structure. Such call path profiles are constructed by overlaying `hpcrun`'s dynamic contexts with static source code structure computed post mortem by the `hpcstruct` tool. In an off-line process called *recovering program structure*, this tool constructs an object to source code mapping. As a result, HPCTOOLKIT's call path profiles expose loops and inlined frames and attribute metrics to them—all for an average overhead of 1–5 % [23].

2.3 Pinpointing Scaling Bottlenecks

We now present an elegant and powerful way to apply HPCTOOLKIT's sampling-based call path profiles to identify scalability bottlenecks in Single-Program Multiple-Data (SPMD) applications, whether executed on a multicore node or a leadership-class supercomputer. The basic idea is to compute a metric that quantifies scaling loss by scaling and differencing call path profiles from a pair of executions [5, 26].

Consider two parallel executions of an application, one executed on p processors and the second executed on $q > p$ processors. In a weak scaling scenario, processors in each execution compute on the same size data. If the application exhibits perfect weak scaling, then the execution times should be identical on both p and q processors. In fact, if every part of the application scales uniformly, then this equality should hold in each corresponding part of the application.

Using HPCTOOLKIT, we collect call path profiles on each of p and q processors to measure the cost associated with each calling context in each execution. HPCTOOLKIT's `hpcrun` profiler uses a data structure called a *calling context tree*

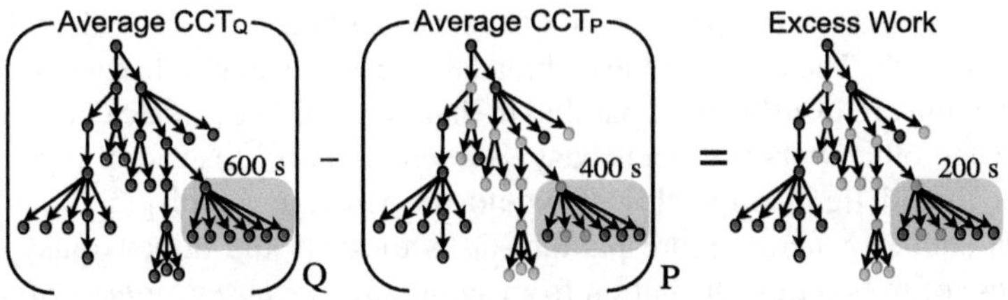

Fig. 2.1 Differencing call path profiles to pinpoint scaling bottlenecks

(CCT) to record a call path profile. Each node in a CCT is identified by a code address. In a CCT, the path from any node to the root represents a calling context. Each node has a weight $w \geq 0$ indicating the exclusive cost attributed to the path from that node to the root. Given a pair of CCTs, one collected on p processors and another collected on q processors, with perfect weak scaling, the cost attributed to all pairs of corresponding CCT nodes should be identical. Any additional cost for a CCT node on q processors when compared to its counterpart in a CCT for an execution on p processors represents *excess work*. This process is shown pictorially in Fig. 2.1. The fraction of excess work, i.e., the amount of excess work in a calling context in a q-process execution divided by the total amount of work in a q-process execution represents the scalability loss attributed to that calling context. By scaling the costs attributed in a CCT before differencing them to compute excess work, one can also use this strategy to pinpoint and quantify strong scalability losses [5]. As long as the CCTs are expected to be similar, this analysis strategy is independent of the programming model and bottleneck cause.

To enable this simple but powerful scalability analysis on the largest class of supercomputers—and to pave the way for other sampling-based tools such as Open|SpeedShop [20] and CrayPat [6]—the HPCTOOLKIT team engaged kernel developers at IBM and Cray to address shortcomings in their microkernels that prevented any sort of sampling-based measurement [26]. We have since applied our scalability analysis to pinpoint synchronization, communication, and I/O bottlenecks in applications on several large-scale distributed-memory machines. Because our analysis is based on differencing comparable executions, we have also used it to pinpoint and quantify scaling bottlenecks on multicore nodes at the loop level [25].

2.4 Blame Shifting

Standard sampling-based measurement identifies symptoms of performance losses rather than causes. Although some symptoms may have an obvious cause, simply knowing symptoms is frequently not sufficient to justify a concrete set of actions for resolving a bottleneck. For example, consider the case of a parallel run time system with idle worker threads. A standard call path profile will show that the idle threads spend a large percentage of their execution time in a scheduling loop waiting for work. The cause of these threads' idleness is that at least some parts of the application are insufficiently parallel in that they fail to generate enough parallel tasks to keep other worker threads busy. But which parts? Instead of reporting that a worker thread frequently idles, it would be better if a tool pinpointed source code that should generate more parallel tasks to keep idle threads busy. We call the process of redirecting attribution from symptoms to causes *blame shifting*.

In general, blame shifting uses either in situ or post mortem analysis to shift the attribution of metrics from performance symptoms to their causes. The need for blame shifting is patently obvious because the primary goal of performance tuning is to find and resolve bottlenecks. However, it is a challenge to make both precise and accurate causal inferences while inducing minimal measurement perturbation. This section discusses three types of blame shifting employed by HPCTOOLKIT.

2.4.1 Parallel Idleness and Overhead in Work Stealing

Work stealing is a popular scheduling technique for dynamic load balancing. One of the most influential versions of work stealing was implemented to support the Cilk language [8] for shared-memory parallel computing. Cilk—along with its descendants Cilk++ and Cilk Plus—combines lazy parallel task creation and a work-stealing scheduler. The Cilk run time maps logically independent asynchronous calls (tasks) onto worker threads (compute cores) in an efficient way. For a Cilk program to execute efficiently, an application must generate enough "well sized" parallel tasks. That is, there should be enough parallel tasks to keep worker threads busy, but those tasks should be large enough so that the overhead of packaging a task (i.e., creating continuations) is insignificant.

The Cilk model challenges existing performance tools. To diagnose causes—not symptoms—of inefficient execution, a tool must solve at least three problems. First, a tool must be able to not only detect when worker threads are idling but also identify the portion of the application causing idleness. Second, a tool must be able to detect when worker threads, though busy, are executing useless work. Third, a tool must be able to attribute its conclusions to source code in its calling context. Although this last challenge may seem trivial, associating costs with the context in which they are incurred is not as simple as it sounds. With a work stealing scheduler, call paths become fragmented because procedure frames migrate between worker threads to balance the work load. Consider a case where thread y steals a task from thread x. In this case, the procedure frames for a source-level call path become separated in both space and time: space, because thread x contains y's parent context; time, because thread x continues executing rather than blocking and waiting for thread y to complete the asynchronous call. As a result, a standard call path profile of a Cilk program yields a result far from what an application developer would expect.

Our solutions to these problems build upon HPCTOOLKIT's call path profiling. To attribute metrics to source-level call paths, we developed logical call path profiling [22, 24], a generalization of call path profiling that maps system-level to source-level call paths. To form the source-level call paths, we stitch together path fragments from a thread's stack and heap data structures.

To identify the causes of insufficient parallelism in Cilk applications, it is necessary to establish, with minimal measurement overhead, a causal link between an idling thread and a working thread that does not generate parallel work. We establish likely suspects using the following scheme [22, 24]. First we make a slight adjustment to the Cilk run-time to always maintain W and I, the number of working and idle threads, respectively. This can be done by maintaining a node-wide counter representing W; since the number of worker threads T is constant we have $I = T - W$. Second, we slightly modify our sampling strategy. If a sample event occurs in a thread that is not working, we ignore it. When a sample event occurs in a thread that is actively working, the thread attributes one sample to the work metric for its sample context. It then obtains W and I and attributes a fractional sample I/W to the idleness metric for the sample context. Even though the thread itself is not idle, it is critical to understand what work the thread is performing when other threads are idle. Our strategy charges the thread its proportional responsibility for not keeping the idle processors busy at that moment at that point in the program. As an example, consider taking a sample of a Cilk execution where five threads are working and three threads are idle. Each working thread records one sample of work in its work metric, and $3/5$ sample of idleness in its idleness metric. The total amount of work and idleness charged for the sample is 5 and 3, respectively.

To identify parallel overhead in Cilk applications, we must distinguish between useful work and overhead. Our solution is based on the observation that it is possible to use static analysis to classify object code instructions as either useful work or overhead. Thus, HPCTOOLKIT can attribute a precise measure of parallel overhead to source-level calling contexts for no additional measurement overhead other than that induced by logical call path profiling [22, 24].

2.4.2 Lock Contention

Locking is the most common means of coordinating accesses while reading and writing mutable shared data structures in threaded programs. Thus, it is important that a performance tool pinpoint causes of lock contention.

There are at least two critical differences between performance monitoring for locks and for Cilk. First, identifying the cause of lock contention requires a more precise causal connection than with idleness in Cilk. With Cilk it is sufficient to observe that at any given instant when threads are idle, all working threads can be thought of as *equally* culpable for not generating enough parallel work to keep idle threads busy. In contrast, with locking, any number of threads may be idling because exactly one working thread holds a lock. Thus, it makes no sense to blame non-lock-holders for idleness. Second, with many lock-based programs, it is not tenable to atomically increment and decrement shared counters on every state change within a thread. In a work stealing run time, steals are relatively infrequent. In contrast, locks may be acquired and released very frequently—and extending key critical sections with an atomic increment could introduce new bottlenecks. Because of these differences, the techniques we described for work stealing are insufficient to offer insight into lock contention. With locks, one must establish causal connections without significantly increasing critical sections and introducing a new bottleneck.

To solve both problems mentioned above, we use locks themselves as a communication channel for precise attribution [28]. For now, we limit our attention to spin (non-blocking) locks. Our technique is based on three rules. First, when a working thread receives a sample, we increment a thread-local work metric. Second, when we sample an idle thread spin-waiting for a lock, we (atomically) increment an idleness metric associated with the lock. The expected number of samples received for all threads waiting on that particular lock is proportional to the time spent waiting for that lock. Furthermore, because the expected number of samples is relatively small when compared with the total number of machine instructions executed during a long-latency spin-wait, the atomic increments are relatively inexpensive. Third, when a working thread releases a lock, we associate all idleness with the lock release point in its full calling context (i.e., as the lock holder releases the lock, it atomically swaps the idleness counter with 0). In this way, we precisely blame the lock release point and accurately attribute to it the idleness that it caused. An important observation is that it is usually straightforward to identify lock acquire points from release points.

We applied our techniques to the MADNESS quantum chemistry application. For the particular input we studied, MADNESS represented a significant challenge: it allocated a total of 65M distinct locks (which were used to implement futures), had a maximum of 340K live locks at any particular instance in the execution; and issued an average of 30K lock acquisitions per second per thread. For a monitoring overhead of a few percent, we found that the primary source of lock contention was in adding work (futures) to a shared work queue [28].

2.4.3 Load Imbalance

Load imbalance is an important problem in programs based on bulk synchronous parallelism. Moreover, many load imbalance problems only appear in medium- to large-scale executions. Even though HPCTOOLKIT can collect detailed call path profiles in a scalable fashion, with profiles it is often difficult to pinpoint load imbalance. In contrast, with a trace (which collects performance information with respect to time) it is relatively easy to identify the effects and possible causes of load imbalance [31]. However, there are significant challenges to collecting fine-grained large-scale traces in a scalable fashion.

Exploiting HPCTOOLKIT's ability to collect large-scale call path profiles, we developed a post mortem analysis that identifies regions of a call path profile that cause load imbalance. Our analysis is based on several observations. First, it is possible to divide an execution into two components: work and exposed (non-sleep) idleness. Second, idleness that occurs around synchronization points is a symptom of load imbalance. Third, the causes of load imbalance are typically reflected in work variability over processes; and for a given procedure or source line in a call path profile, one can compute statistics over all processes in an execution that measure variability. Fourth, with layered software, important imbalance and synchronization points are often context sensitive. For example, an application may have many calls to `MPI_Allreduce`. While the work between some pairs of `MPI_Allreduce` operations may be balanced, between other pairs it may not, causing lightly-loaded threads to idle. Given these observations, we developed a post mortem blame shifting analysis of call path profiles that identifies exposed idleness and associates that idleness with possible causes [27].

2.5 Call Path Tracing

Although tracing is a powerful performance-analysis technique, tools that employ it can quickly become bottlenecks themselves. To avoid large overheads, instrumentation-based tracing tools typically monitor only a subset of the procedures in an application. However, even coarse-grained trace data of large-scale executions quickly becomes unwieldy and a challenge to present. Unfortunately, to obtain actionable performance feedback for modular parallel software systems, it is often necessary to have fine-grained context-sensitive data—the very thing scalable tools avoid. We have developed sampling-based techniques to collect and present arbitrary slices of fine-grained large-scale call path traces [29].

Figure 2.2 shows a screen shot of HPCTOOLKIT's presentation tool `hpctrace viewer`. As with other tools that present space-time visualizations, time advances from left to right on a horizontal axis and MPI-rank space advances from top to bottom on a vertical axis. Unlike other tools, `hpctraceviewer`'s renderings are hierarchical. Since each sample in each process's time line represents a call path, we can view the process timelines at different call path depths to show more or less detailed views of the execution. The Figure shows one slice of the process/time/call-path space at depth 3; the inset shows the selected region at depths 3, 6 and 7.

To collect traces, HPCTOOLKIT exploits a useful property of our CCT data structure: a full call path can be represented by a single leaf node. Thus, we can generate a full call path trace by maintaining a CCT and stream of compact trace records where each trace record is 12 bytes (4-byte CCT node id and an 8-byte timestamp). For reasonable sampling rates, the data rate of this trace stream is extremely low. Moreover, because sampling rates are controllable, the granularity of the trace and its corresponding data volume can scale to very large or long executions.

To present traces, `hpctraceviewer` also uses sampling. Given a display window of height h and width w (in pixels), the tool simply samples the trace data records to determine how each pixel should be colored. Using sampling,

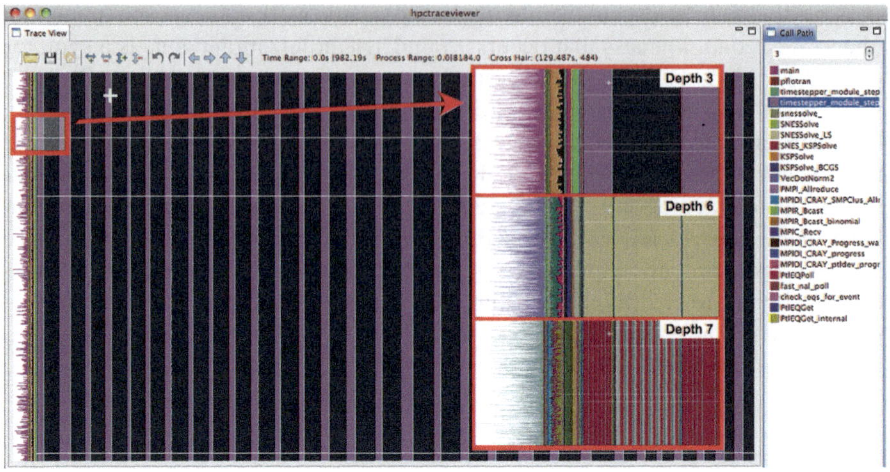

Fig. 2.2 An 8184-core execution of PFLOTRAN on a Cray XT5. The inset exposes call path hierarchy by showing the selected region (*top left*) at different call path depths

`hpctraceviewer` can render call-path depth slices at arbitrary resolutions in time $O(hw \log t)$, where t is the number of trace records in the largest trace file. In practice, the $\log t$ factor is often negligible because we use linear extrapolation to predict the locations of the w trace records. The practical upshot of this technique is that `hpctraceviewer` can rapidly present trace files of arbitrary length for executions with an arbitrary number of threads on a laptop.

2.6 Data-Centric Performance Analysis

Sampling-based profilers such as HPCTOOLKIT attribute samples to object and source code for very low overhead. By exploiting recent hardware support, HPC-TOOLKIT uses sampling to attribute measurements not only to source code but also to data objects [13]. With this *data-centric* attribution, HPCTOOLKIT can attribute metrics such as memory access latency to program data objects and their corresponding source code use points—in its full calling context. The results are especially useful when long-lived data objects are accessed at many places.

2.7 Conclusions

We have found that sampling-based techniques are surprisingly versatile and effective. Using sampling-based measurement, HPCTOOLKIT can collect exquisitely detailed call path profiles and traces for an average run-time overhead of less than 5 %. Using sampling-based presentation, HPCTOOLKIT effortlessly presents arbitrary slices of large-scale fine-grained call path traces. To make sampling-based call path profiling of fully optimized applications possible, HPCTOOLKIT employs a set of innovative techniques to collect call paths at arbitrary points in an execution (using stack unwinding) and attribute to loops and inlined functions. In many cases HPCTOOLKIT's code-centric profiles are sufficient for understanding performance problems, but for more difficult problems, it is necessary to understand how aggregate costs arise over time or with respect to data. To pinpoint and quantify many scaling bottlenecks, it is sufficient to simply compare two or more call path profiles—accurate call path profiles that show statements in their full calling context and expose loops. To offer more insight into the *causes* of scaling bottlenecks, we developed several problem-focused analyses that shift blame from a bottleneck's effects to its causes or potential causes. Because they are based on sampling, these techniques incur very low measurement overhead and apply to problems as diverse as lock contention, load imbalance, insufficient parallelism in work stealing, and parallel overhead. The ability to collect and present accurate, detailed and problem-focused measurements for large-scale production executions has enabled HPCTOOLKIT's use on today's grand challenge applications: multi-lingual programs leveraging third-party libraries for which source code and symbol information are often unavailable.

In many ways, sampling is a perfect fit for performance analysis. One issue with sampling-based methods is that while they tend to show representative behavior, they might miss interesting extreme behavior. While this property may limit the use of sampling in correctness tools, we have not found it to be problematic for performance tools. The reason is that, assuming an appropriately sized and uncorrelated sample, sampling-based methods expose anomalies *that generally affect an execution*. In other words, although using sampling-based methods can miss any specific anomaly, those same methods will expose the anomaly's effects if they are important for performance. If the anomaly does not interfere with the application's execution, then it is not particularly important with respect to performance. If on the other hand, process synchronization transfers the anomaly's effects to other process ranks, we expect sampling-based methods to expose that fact.

Acknowledgements HPCTOOLKIT would not be what it is without the efforts of Mark Krentel, Laksono Adhianto, and Mike Fagan. Xu Liu developed our data-centric analysis.

References

1. Adhianto, L., Banerjee, S., Fagan, M., Krentel, M., Marin, G., Mellor-Crummey, J., Tallent, N.R.: HPCToolkit: Tools for performance analysis of optimized parallel programs. Concurr. Comput. Pract. Exp. **22**(6), 685–701 (2010)
2. Adhianto, L., Mellor-Crummey, J., Tallent, N.R.: Effectively presenting call path profiles of application performance. In: International Conference on Parallel Processing Workshops, pp. 179–188. IEEE Computer Society, Los Alamitos (2010)

3. Arnold, M., Ryder, B.G.: A framework for reducing the cost of instrumented code. In: Proceedings of the 2001 ACM SIGPLAN Conference on Programming Language Design and Implementation, pp. 168–179. ACM, New York (2001)
4. Chung, I.H., Walkup, R.E., Wen, H.F., Yu, H.: MPI performance analysis tools on Blue Gene/L. In: Proceedings of the 2006 ACM/IEEE Conference on Supercomputing, p. 123. ACM, New York (2006)
5. Coarfa, C., Mellor-Crummey, J., Froyd, N., Dotsenko, Y.: Scalability analysis of SPMD codes using expectations. In: Proceedings of the 21st International Conference on Supercomputing, pp. 13–22. ACM, New York (2007)
6. De Rose, L., Homer, B., Johnson, D., Kaufmann, S., Poxon, H.: Cray performance analysis tools. In: Tools for High Performance Computing, pp. 191–199. Springer, Berlin (2008)
7. Free Software Foundation: Glibc. http://www.gnu.org/s/libc/ (2012)
8. Frigo, M., Leiserson, C.E., Randall, K.H.: The implementation of the Cilk-5 multithreaded language. In: Proceedings of the 1998 ACM SIGPLAN Conference on Programming Language Design and Implementation, pp. 212–223. ACM, New York (1998)
9. Froyd, N., Mellor-Crummey, J., Fowler, R.: Low-overhead call path profiling of unmodified, optimized code. In: Proceedings of the 19th International Conference on Supercomputing, pp. 81–90. ACM, New York (2005)
10. Geimer, M., Wolf, F., Wylie, B.J.N., Ábrahám, E., Becker, D., Mohr, B.: The Scalasca performance toolset architecture. Concurr. Comput. Pract. Exp. **22**(6), 702–719 (2010)
11. Hollingsworth, J.K., Miller, B.P., Cargille, J.: Dynamic program instrumentation for scalable performance tools. In: Proceedings of the 1994 Scalable High Performance Computing Conference, pp. 841–850. IEEE Computer Society, Los Alamitos, CA, USA (1994)
12. Knüpfer, A., Brunst, H., Doleschal, J., Jurenz, M., Lieber, M., Mickler, H., Müller, M.S., Nagel, W.E.: The Vampir performance analysis tool-set. In: Resch, M., Keller, R., Himmler, V., Krammer, B., Schulz, A. (eds.) Tools for High Performance Computing, pp. 139–155. Springer, Berlin (2008)
13. Liu, X., Mellor-Crummey, J.: Pinpointing data locality problems using data-centric analysis. In: Proceedings of the 2011 IEEE/ACM International Symposium on Code Generation and Optimization, Chamonix, France, pp. 171–180. IEEE Computer Society, Los Alamitos (2011)
14. Malony, A.D., Shende, S., Morris, A., Wolf, F.: Compensation of measurement overhead in parallel performance profiling. Int. J. High Perform. Comput. Appl. **21**(2), 174–194 (2007)
15. Mellor-Crummey, J., Fowler, R., Marin, G., Tallent, N.: HPCView: a tool for top-down analysis of node performance. J. Supercomput. **23**(1), 81–104 (2002)
16. Miller, B.P., Callaghan, M.D., Cargille, J.M., Hollingsworth, J.K., Irvin, R.B., Karavanic, K.L., Kunchithapadam, K., Newhall, T.: The Paradyn parallel performance measurement tool. Computer **28**(11), 37–46 (1995)
17. Mosberger-Tang, D.: libunwind. http://www.nongnu.org/libunwind (2012)
18. Petrini, F., Kerbyson, D.J., Pakin, S.: The case of the missing supercomputer performance: achieving optimal performance on the 8,192 processors of ASCI Q. In: Proceedings of the 2003 ACM/IEEE Conference on Supercomputing, p. 55. IEEE Computer Society, Washington, DC (2003)
19. Rice University: HPCToolkit performance tools. http://hpctoolkit.org (2012)
20. Schulz, M., Galarowicz, J., Maghrak, D., Hachfeld, W., Montoya, D., Cranford, S.: Open|SpeedShop: an open source infrastructure for parallel performance analysis. Sci. Program. **16**(2–3), 105–121 (2008)
21. Shende, S.S., Malony, A.D.: The TAU parallel performance system. Int. J. High Perform. Comput. Appl. **20**(2), 287–311 (2006)
22. Tallent, N.R., Mellor-Crummey, J.: Effective performance measurement and analysis of multithreaded applications. In: Proceedings of the 14th ACM SIGPLAN Symposium on Principles and Practice of Parallel Programming, pp. 229–240. ACM, New York (2009)
23. Tallent, N.R., Mellor-Crummey, J., Fagan, M.W.: Binary analysis for measurement and attribution of program performance. In: Proceedings of the 2009 ACM SIGPLAN Conference on Programming Language Design and Implementation, pp. 441–452. ACM, New York (2009)

24. Tallent, N.R., Mellor-Crummey, J.M.: Identifying performance bottlenecks in work-stealing computations. Computer **42**(12), 44–50 (2009)
25. Tallent, N., Mellor-Crummey, J., Adhianto, L., Fagan, M., Krentel, M.: HPCToolkit: performance tools for scientific computing. J. Phys. Conf. Ser. **125**, 012088 (5pp) (2008)
26. Tallent, N.R., Mellor-Crummey, J.M., Adhianto, L., Fagan, M.W., Krentel, M.: Diagnosing performance bottlenecks in emerging petascale applications. In: Proceedings of the 2009 ACM/IEEE Conference on Supercomputing, pp. 1–11. ACM, New York (2009)
27. Tallent, N.R., Adhianto, L., Mellor-Crummey, J.M.: Scalable identification of load imbalance in parallel executions using call path profiles. In: Proceedings of the 2010 ACM/IEEE Conference on Supercomputing, pp. 1–11. IEEE Computer Society, Washington, DC (2010)
28. Tallent, N.R., Mellor-Crummey, J.M., Porterfield, A.: Analyzing lock contention in multi-threaded applications. In: Proceedings of the 15th ACM SIGPLAN Symposium on Principles and Practice of Parallel Programming, pp. 269–280. ACM, New York (2010)
29. Tallent, N.R., Mellor-Crummey, J.M., Franco, M., Landrum, R., Adhianto, L.: Scalable fine-grained call path tracing. In: Proceedings of the 25th International Conference on Supercomputing, pp. 63–74. ACM, New York (2011)
30. Traub, O., Schechter, S., Smith, M.D.: Ephemeral instrumentation for lightweight program profiling. Tech. rep., Harvard University (1999)
31. Wolf, F., Wylie, B.J.N., Ábrahám, E., Becker, D., Frings, W., Fürlinger, K., Geimer, M., Hermanns, M.A., Mohr, B., Moore, S., Pfeifer, M., Szebenyi, Z.: Usage of the Scalasca toolset for scalable performance analysis of large-scale parallel applications. In: Tools for High Performance Computing, pp. 157–167. Springer, Berlin (2008)

Chapter 3
likwid-bench: An Extensible Microbenchmarking Platform for x86 Multicore Compute Nodes

Jan Treibig, Georg Hager, and Gerhard Wellein

Abstract Microbenchmarking is an essential tool for characterizing modern compute nodes. Apart from determining raw performance capabilities microbenchmarking can be used to aquire input parameters for performance models or mimic the behavior of more complex applications. Many existing microbenchmarks are not extensible and implemented in C or Fortran. One problem with microbenchmarks in a high level language is that many performance issues are only apparent on the instruction level. The code quality of the compiler is an additional source of variation. likwid-bench is a framework enabling rapid prototyping of loop-based, threaded assembly kernels. It eases the process of implementing assembly kernels by providing a portable assembly language independent from any concrete assembler program. likwid-bench already includes many standard microbenchmarking testcases and can be used out of the box as a microbenchmarking tool.

3.1 Introduction

Microbenchmarking is an essential tool to investigate the interaction of software with the hardware. It can be used to determine upper limits of performance characteristics on compute nodes as well as pin down anomalies in the microarchitecture of a processor. Another important application is to acquire input parameters of performance models. A prominent example for a microbenchmark is the STREAM benchmark proposed by McCalpin [6], which is the standard for sustained main memory bandwidth and is often used as baseline for the balance metric approach [10]. The theoretical performance numbers published by the hardware

J. Treibig (✉) · G. Hager · G. Wellein
Erlangen Regional Computing Center (RRZE), Friedrich-Alexander Universität Erlangen-Nürnberg, Martensstr. 1, D-91058, Erlangen, Germany
e-mail: jan.treibig@rrze.uni-erlangen.de; georg.hager@rrze.uni-erlangen.de; gerhard.wellein@rrze.uni-erlangen.de

H. Brunst et al. (eds.), *Tools for High Performance Computing 2011*,
DOI 10.1007/978-3-642-31476-6_3,

vendors do often not reflect the performance achievable by real applications. Since instruction code has a large impact on performance, a high level language such as C is not suited to determine these numbers because it introduces another layer of abstraction depending on compiler quality. This is even more important in the x86 world were new instruction set extensions are released with high frequency. In [8] the Open Source LIKWID tool suite [2] was introduced as a set of lightweight performance-related command line tools for the Linux operating system. In [9] new functionality and application scenarios were described, including the benchmarking application likwid-bench. In this paper we want to cover likwid-bench in more detail and elaborate on implementation questions. likwid-bench is a microbenchmarking framework allowing rapid prototyping of stream based, threaded assembly kernels. Because it comes with a set of common microbenchmarking kernels it can be used out of the box as a benchmarking application. It is also a valuable tool for measuring instruction code performance issues for code generators or compilers.

Implementing an assembly kernel with consistent time measurement is non-trivial. On multicore chips threading is required to determine issues like resource contention or bottlenecks. Complex node topologies require strict thread core affinity to get reliable performance characteristics. Finally data placement issues due to ccNUMA effects have to be taken into account.

likwid-bench provides a framework which allows the implementer to concentrate on the instruction code of the benchmarking kernel. The high level assembly language provided eases this process as far as possible without hiding any important aspect. The runtime cares for threading and affinity, for data allocation and placement, for accurate time measurement, and for result presentation. All necessary configuration for benchmark execution is passed on the command line.

This paper is organized as follows. In Sect. 3.3 the software architecture of likwid-bench is described. The special benchmark text file format used to create new benchmarks is introduced in Sect. 3.4. Section 3.6 shows application examples, and Sect. 3.7 presents conclusion and an outlook to future work.

3.2 Related Work

About a decade ago existed benchmarking tools for low level measurement of latency and bandwidth of the memory hierarchy on modern cache based processors especially in the UNIX environment. Two prominent examples are LLCBench [4] and lmbench [5]. In [1] the evaluation of hardware performance counter data acquired with PAPI using microbenchmarking is described. In [11] a tool is described to automatically determine hardware parameters later used for automatic performance optimization. That paper puts much emphasis on the use of C as a portable "assembly language." The IPACS project [3] aimed at providing a unified framework for various benchmarking backends and result presentation.

likwid-bench ultimately tries to provide the user with enough insight into the microarchitecture to understand the interaction of application code with the hardware on the lowest sensible level. It does not use C for benchmarking kernels

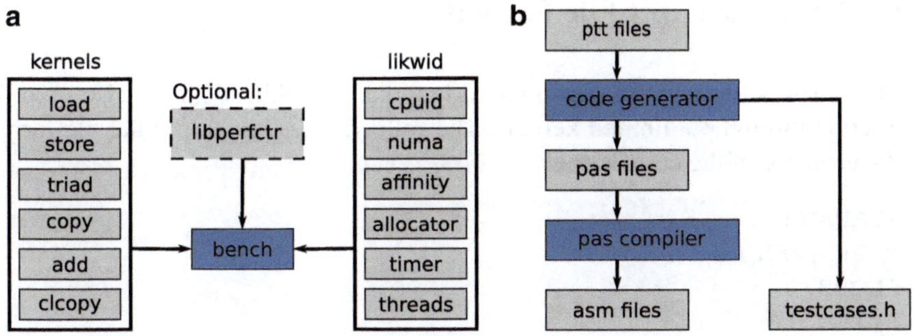

Fig. 3.1 Software architecture of likwid-bench. (**a**) Architecture. (**b**) Benchmark generation

because the exact instruction code has a crucial influence on performance metrics. We do not know of any project providing a user configurable runtime environment for threaded assembly kernels with the same flexibility in thread and data placement.

3.3 Architecture

likwid-bench is builds on the components of the LIKWID toolsuite. It uses the cpuid module to extract topology information and the affinity module to pin threads. The allocator module manages aligned data allocation and placement using the ccNUMA information provided by the numa module. For the threaded execution the threads module provides a flexible interface with thread groups and fast synchronization using the `pthreads` API. Optionally likwid-bench can be configured to use the libperfctr API instrumentation calls providing very accurate performance counter measurements. The overall architecture of likwid-bench is illustrated in Fig. 3.1.

The benchmarking kernels are text files in the .ptt file format, describing the loop body instruction code accompanied by meta information necessary for the execution and result presentation. Benchmarks are named according to the ptt file names. During the build process a perl script reads in all .ptt files in a specific directory and converts them to a high level assembly file format (.pas) and C header with prototype declarations of all testcases. For every testcase an assembly function is generated which is later called from the benchmarking core application.The .pas files are converted to assembler files (currently for the gas assembler) and finally assembled to an object file. The intermediate .pas file provides an abstraction from the assembler format and would allow, e.g., to port likwid-bench to Windows using the Microsoft assembler. All intermediate build products are available allowing to closely review the generated assembly code. To add a new benchmark the user must add a new .ptt file in the bench directory and recompile.

3.4 Benchmark .ptt File Format

likwid-bench is focused on stream based loop kernels. Still it can be also used for instruction throughput limited kernels. The following listing shows the file format on the example of the copy kernel:

```
STREAMS 2
TYPE DOUBLE
FLOPS 0
BYTES 16
LOOP 8
movaps     FPR1, [STR0 + GPR1 * 8]
movaps     FPR2, [STR0 + GPR1 * 8 + 16]
movaps     FPR3, [STR0 + GPR1 * 8 + 32]
movaps     FPR4, [STR0 + GPR1 * 8 + 48]
movaps     [STR1 + GPR1 * 8]      , FPR1
movaps     [STR1 + GPR1 * 8 + 16], FPR2
movaps     [STR1 + GPR1 * 8 + 32], FPR3
movaps     [STR1 + GPR1 * 8 + 48], FPR4
```

Data is provided by means of vector streams (`STR0`, `STR1`). The necessary number of streams used by the benchmark must be configured accompanied by the data type (at the moment single and double precision floats and integers are supported). For result output it must be further specified how many flops are executed and how many bytes are transfered in a single scalar update operation. The `LOOP` statement marks the beginning of the loop body. The loop control code is automatically generated, `GPR1` being the default loop counter. The number following the `LOOP` statement configures how many scalar updates are performed in one loop body iteration. The loop counter is incremented by this number after the execution of the loop body. In above example the loop body is four-way unrolled. Since the code performs two updates per instruction due to SIMD vectorization, one loop body iteration performs eight updates. The assembly code must use Intel assembler syntax. For convenience a number of placeholders are provided: Before the `LOOP` statement setup instruction code can be placed, e.g. to initialize registers.

`FPR[0-15]`	SIMD vector registers
`GPR[0-15]`	General purpose registers
`STR[0-10]`	Registers with stream addresses
`SCALAR`	Vector double precision constant
`SSCALAR`	Vector single precision constant
`ISCALAR`	Vector integer constant

3.5 Command Line Syntax

The command line syntax provides a simple and intuitive but yet powerful interface for thread and data placement. likwid-bench uses the notion of *thread groups*. A thread group is a subset of threads operating on the same data set. For every thread group a workgroup command line option must be present describing the thread and optionally data placement. For thread and data placement the notion of so called *thread domains* introduced is used. A thread domain is an entity on the node shared by a thread group. The top level thread domain is the node, followed by ccNUMA locality domains, sockets, and shared caches. If there are multiple domains of the same type they are numbered consecutively starting with zero. Thread domains are specified using single characters. This allows a simple interface accounting for all current and also future topology developments. At the moment the following thread domains are supported:

`N`	Node
`M[0-N]`	ccNUMA locality domain
`S[0-N]`	Sockets
`C[0-N]`	Last level shared cache

Required information for defining a work group is the thread domain to be used and the size of the working set. This size is the total size used for all streams in a benchmark. The necessary vector lengths for each stream is calculated by the environment (alignment and padding issues are handled automatically). The simplest invocation for likwid-bench using the copy example is:

```
$ likwid-bench -t copy -i 1000 -g 1 -w S0:2MB
```

The `-t` parameter specifies the benchmark type and `-g` allows to set the number of thread groups. At the moment the number of iterations must be given on the command line with the `-i` option. It is planned to make this an optional information with the default ensuring a runtime sufficient to get exact results. The `-w` option configures one thread group. Reasonable defaults are used for all omitted information. In the example above the default is to use all threads available in the thread domain, including logical cores. The following invocation will use a limited number of four threads:

```
$ likwid-bench -t copy -i 1000 -g 1 -w S0:2MB:4
```

Every successive thread is pinned to distinct cores. This means that for, e.g., an Intel Quad core Nehalem processor four threads will be pinned to all physical cores. If six threads are specified the physical cores will be used first and the remaining two threads will be pinned to the first two logical cores. This simple approach of filling up the thread domains has one disadvantage: It does not support more sophisticated thread placement policies. This is apparent on the upcoming AMD "Interlagos" processor: Two cores share one floating point unit. With the current approach it is not possible, e.g., to skip every other core to only use one core per floating point unit. We plan to extend the current syntax to allow strides without complicating the basic approach.

For data placement the default is that all data is placed in the same thread domain the threads are running in. Optionally the exact placement for every stream used can be configured. The copy benchmark has two streams; this value can be queried from likwid-bench with the command `likwid-bench -l <testcase>`. The following command:

```
$ likwid-bench -t copy -i 1000 -g 1 \
  -w S0:2MB:4-0:S1,1:S1
```

will place both streams (`0` and `1`) in the threading domain socket 1 (`S1`). Each stream is page aligned per default. Optionally an offset may be specified separately for each stream.

3.6 Examples

In the following we present possible applications of likwid-bench. The examples demonstrate critical issues of modern multicore compute nodes: Memory and last level cache bandwidth saturation, and ccNUMA characteristics.

3.6.1 Identifying Bandwidth Bottlenecks

Determining bandwidth bottlenecks on a compute node is critical for judging application performance. If it is known what bottlenecks are apparent on a node, performance limitations can be better understood and techniques can be applied to work around those bottlenecks. This applies to main memory but also to shared last level cache bandwidths.

likwid-bench provides special cacheline wise variants of basic data operations such as load, store, and copy. Because in our case the raw capabilities of the memory subsystem are needed the runtime spent in executing instructions must be kept at a minimum [7]. This is ensured by only touching one data item per cacheline, which forces each cacheline into the L1 cache but minimizes the time spent on instruction execution. These benchmark variants show the peak cache bandwidths achievable by software. Figure 3.2a shows last level cache bandwidth scalability when increasing the number of threads. Clearly the Intel Westmere processor shows a bandwidth bottleneck starting at three threads. It can also be seen that while for the load operation the processor is able to saturate its peak cache bandwidth, contention occurs for the copy operation leading to roughly half of the possible peak bandwidth. The successor SandyBridge solves both issues, since it incorporates a cache design which uses multiple segments connected by a segmented ring bus. This makes the

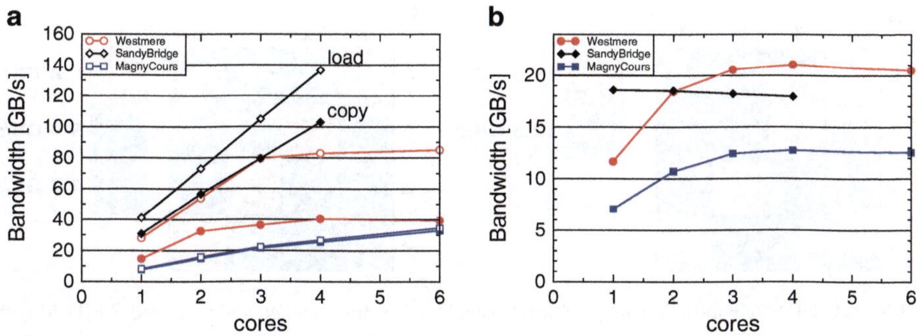

Fig. 3.2 Bandwidth saturation of shared last level caches (**a**) and main memory (**b**). In both cases special cacheline wise versions of load and copy (**a**) or copy (**b**) are used (Intel Westmere Xeon X5670, Intel SandyBridge Core i7-2600 and a AMD MagnyCours Opteron 6176SE)

L3 cache a parallel device and provides scaling L3 bandwidth. Also the difference in performance between load and copy is much smaller.

For AMD MagnyCours a six core element as the largest entity with a shared cache is used for comparison. The L3 cache on MagnyCours is not well suited for bandwidth-demanding operations. It scales well up to six cores, but on a low absolute performance level. Up to three cores the bandwidth is not significantly larger than main memory bandwidth. Still on the full socket it can draw level with Intel Westmere. There is no difference between load and copy, which can be attributed to the write allocate exclusive cache design, triggering the same number of cacheline transfers for load and copy.

Figure 3.2b shows memory bandwidth scaling inside one ccNUMA locality domain. Both Intel Westmere and AMD MagnyCours show a bandwidth saturation starting with three threads. The desktop SandyBridge processor is able to saturate its bandwidth with a single thread.

3.6.2 Characterizing ccNUMA Properties

Cache coherent non-uniform memory architectures are used today to provide scalable memory bandwidth performance on the node level. This comes at the price of an increased complexity for the programmer with regard to performance. While on current Intel machines the ccNUMA locality domain is equivalent to the socket, on recent AMD processors one physical socket is made up of two dies eachg with its own locality domains. This means that on a four socket AMD MagnyCours system eight locality domains exist connected by the Hypertransport bus. To obey ccNUMA locality is therefore crucial to get good performance on such nodes. With likwid-bench it is easy to mimic application characteristics with regard to ccNUMA

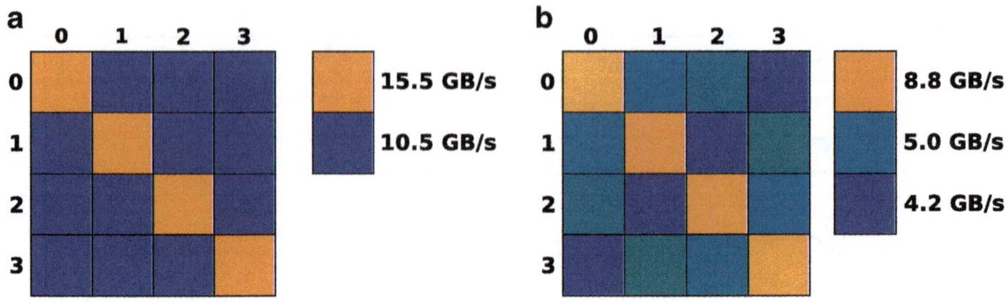

Fig. 3.3 NUMA bandwidth map of Intel Nehalem EX four socket node (**a**) and AMD Magny Cours two socket node with four ccNUMA locality domains. The Testcase is a standard memory copy operation with point to point transfer between two locality domains. All cores of a locality domain are used. Bencwidth result whoen are effective bandwidth seen by the application, i.e., not counting write allocate transfers. (**a**) Nehalem EX 4S. (**b**) Magny Cours 2S

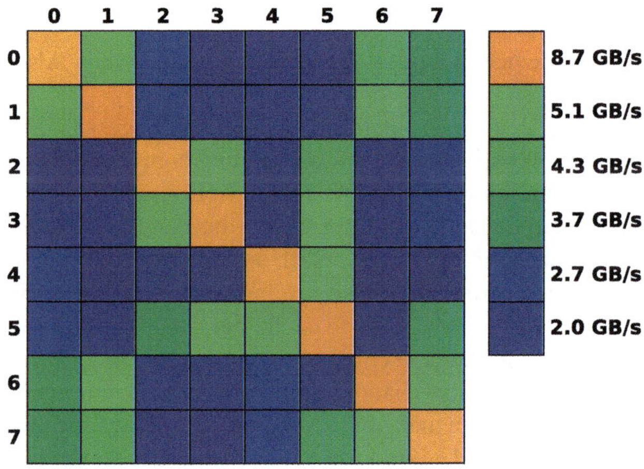

Fig. 3.4 NUMA bandwidth map of four socket AMD MagnyCours system using a standard memory copy operation

performance. In recent releases of the likwid tool suite a small perl script is included which automatically creates a graphical ccNUMA bandwidth map of a complete node based on a memory copy operation.

Figure 3.3a shows the NUMA bandwidth map for an Intel Nehalem EX four socket node. Surprisingly it has only two performance domains: Local with 15.5 GB/s and remote with 10.5 GB/s bandwidth. On the two socket MagnyCours node with four locality domains three performance domains can be distinguished. Still the difference between the remote domains is negligible. The drop from local to remote is larger on this AMD system than on the Intel Nehalem EX. Figure 3.4 shows the bandwidth map of a four socket AMD MagnyCours system with eight locality domains. This system shows a strong NUMA characteristic with a bandwidth drop for remote accesses to only 25 % of the local bandwidth.There is a large variation between remote accesses of over a factor of 2. The results cannot be fully explained by the Hypertransport topology on the node alone. It is suspected that routing issues influence the results.

3.7 Conclusion and Outlook

likwid-bench is a flexible microbenchmarking platform for streaming loop kernels. It offers great flexibility with regard to thread and data placement and allows to measure many performance-relevant aspects of modern compute nodes. The inherent limitation to a certain family of architectures can also be seen as an advantage because it allows full control over the executed instruction code. We have shown the application of likwid-bench on several examples.

One drawback of the current implementation is that it is restricted to simple stream based data access. To mimic the behavior of more complex algorithms like, e.g., stencil codes multi-dimensional data layouts have to be supported. Also the current syntax specifying thread placement is not flexible enough as it does not allow free placement but is based on filling up thread domains.

One aspect of likwid-bench is to form a community platform for exchanging optimal implementations of low level kernels on different platforms. This point will be in the focus for future activities. There are plans to provide domain specific kernel packages which can be installed and reviewed to get information how to implement algorithmic primitives.

Acknowledgements We are indebted to Intel Germany for providing test systems and early access hardware for benchmarking. This work was supported by the Competence Network for Scientific and Technical High Performance Computing in Bavaria (KONWIHR) under the project "OMI4papps."

References

1. Araiza, R., Pham, T., Aguilera, M.G.: Towards a cross-platform microbenchmark suite for evaluating hardware performance counter data. In: Richard Tapia Celebration of Diversity in Computing Conference, Albuquerque. New Mexico (2005)
2. Homepage of LIKWID tool suite. http://code.google.com/p/likwid/
3. Kredel, H., Merz, M.: The design of the IPACS distributed software architecture. In: 2nd Workshop on Distributed Objects Research, Experiences Applications (DOREA 2), Las Vegas (2004)
4. LLCbench – Low Level Architectural Characterization Benchmark Suite *Homepage*. http://icl.cs.utk.edu/projects/llcbench/
5. LMbench – Tools for Performance Analysis *Homepage*. http://lmbench.sourceforge.net
6. McCalpin, J.D.: Memory bandwidth and machine balance in current high performance computers. In: IEEE Computer Society Technical Committee on Computer Architecture (TCCA) Newsletter. New York (1995)

7. Treibig, J., Hager, G.: Introducing a performance model for bandwidth-limited loop kernels. In: Proceedings of the Workshop Memory issues on Multi- and Manycore Platforms at PPAM 2009, Wroclaw (2009)
8. Treibig, J., Hager, G., Wellein, G.: LIKWID: a lightweight performance-oriented tool suite for x86 multicore environments. In: Proceedings of PSTI2010, the First International Workshop on Parallel Software Tools and Tool Infrastructures, San Diego (2010)
9. Treibig, J., Hager, G., Wellein, G.: LIKWID performance tools. In: Bischof, C., et al. (eds.) Competence in High Performance Computing 2010, pp. 165–175. Springer, ISBN 978-3-642-24025-6 (2012)
10. Williams, S., Waterman, A., Patterson, D.: Roofline: an insightful visual performance model for multicore architectures. Commun. ACM **52**, 65–76 (2009). New York
11. Yotov, K., Pingali, K., Stodghill, P.: X-ray: a tool for automatic measurement of hardware parameters. In: Second International Conference on the Quantitative Evaluation of Systems, Torino. IEEE Computer Society, Los Alamitos (2005)

Chapter 4
An Open-Source Tool-Chain for Performance Analysis

Kevin Coulomb, Augustin Degomme, Mathieu Faverge, and François Trahay

Abstract Modern supercomputers with multi-core nodes enhanced by accelerators as well as hybrid programming models introduce more complexity in modern applications. Efficiently exploiting all of the available resources requires a complex performance analysis of applications in order to detect time-consuming or idle sections. This paper presents an open-source tool-chain for analyzing the performance of parallel applications. It is composed of a trace generation framework called EZTRACE, a generic interface for writing traces in multipe formats called GTG, and a trace visualizer called VITE. These tools cover the main steps of performance analysis – from the instrumentation of applications to the trace analysis – and are designed to maximize the compatibility with other performance analysis tools. Thus, these tools support multiple file formats and are not bound to a particular programming model. The evaluation of these tools show that they provide similar performance compared to other analysis tools.

K. Coulomb (✉)
SysFera, Lyon, France
e-mail: kevin.coulomb@sysfera.com

A. Degomme
INRIA Rhône-Alpes, Grenoble, France
e-mail: augustin.degomme@inrialpes.fr

M. Faverge
Innovative Computing Laboratory, University of Tennessee, Knoxville, TN, USA
e-mail: mfaverge@eecs.utk.edu

F. Trahay
Institut Télécom, Télécom SudParis, Evry Cedex, France
e-mail: francois.trahay@it-sudparis.eu

H. Brunst et al. (eds.), *Tools for High Performance Computing 2011*,
DOI 10.1007/978-3-642-31476-6_4, © Springer-Verlag Berlin Heidelberg 2012

4.1 Introduction

Numerical simulation has become one of the pillars of science in many domains: numerous research topics now rely on computational simulations for modeling physical phenomenons. The need for simulation in various computer power hungry research areas, such as climate modeling, computational fluid dynamics, and astrophysics has led to designing massively parallel computers that now reach petaflops. Given the cost of such supercomputers, high performance applications are designed to exploit the available computing power to its maximum. During the development of an application, the optimization phase is crucial for improving the efficiency. However, this phase requires extensive understanding of the behavior and the performance of the application. The complexity of supercomputer hardware, due to the use of NUMA architectures or hierarchical caches, as well as the use of various programming models like MPI, OpenMP, MPI+threads, MPI+GPUs and PGAS models, makes it more and more difficult to understand the performance of an application. Due to the complexity of the hardware and software stack, the use of convenient analysis tools is a great help for understanding the performance of an application. Such tools permit the user to follow the behavior of a program and to spot its problematic phases.

This paper describes a complete set of tools designed for performance analysis, from the instrumentation of parallel applications using EZTRACE to the analysis of their execution with VITE. This open-source tool-chain provides a convenient and performant means to understand the behavior of an application.

The remainder of this paper is organized as follows: in Sect. 4.2, we present various research related to performance analysis. The design of EZTRACE – our instrumentation framework – is described in Sect. 4.3. Section 4.4 presents the GTG tracing library. Section 4.5 provides an overview of our trace visualization tool named VITE. The results of experiments conducted on EZTRACE are discussed in Sect. 4.6. Finally, in Sect. 4.7, we draw a conclusion and introduce future work.

4.2 Related Work

Since the advent of parallel programming and the need for optimized applications, numerous work has been conducted on performance analysis. Tools were designed for tracing the execution of parallel applications in order to understand their behavior. Some of these tools are specific to a particular programming model – MPE [4] targets MPI applications, POSIX THREAD TOOL [6] aims at applications that use pthreads, OMPTRACE [3] instruments OpenMP applications, ... – Others, such as VAMPIRTRACE [11], TAU [13] or SCALASCA [8], provide multiple modules and thus support multiple programing models. Instrumenting custom API or applications can be achieved with these tools by manually or automatically instrumenting the code.

The format of the trace generated by a tracing tool is usually specific, leading to incompatibility between performance analysis tools. Generic trace formats were designed to meet the needs of several tools. The PAJ format [7] permits the user to depict the execution of a program in a generic and hierarchic way. The OPEN TRACE FORMAT [9] (OTF) provides a generic and scalable means of tracing parallel applications more adapted to MPI applications using various communicators.

Exploring a trace file thus requires a tool designed for a particular trace format. For instance, OTF traces can be viewed with VAMPIR [10], TRIVA [12] displays PAJ traces, and the files generated with MPE can be visualized with JUMPSHOT [4]. TAU and SCALASCA embed their own trace file viewer. The lack of multiformat trace viewers forces users to switch from one system to another, depending on the tracing tool in use. A new complete tool-chain – from the application tracing to the trace analyzer – able to manipulate several trace formats, would allow users to use the most relevant format for each application to analyze while keeping the same tools.

4.3 Instrumenting Applications with EZTRACE

EZTRACE [14] has been designed to provide a simple way to trace parallel applications. This framework relies on *plugins* in order to offer a generic way to analyze programs; depending on the application to analyze or on the point to focus on, several modules can be loaded. EZTRACE provides predefined *plugins* that give the ability to the user to analyze applications that use MPI libraries, OpenMP or Pthreads as well as hybrid applications that mix several of these programming models. However, user-defined *plugins* can also be loaded in order to analyze application functions or custom libraries.

EZTRACE uses a two-phases mechanism for analyzing performance. During the first phase that occurs while the application is executed, functions are intercepted and events are recorded. After the execution of the application, the *post-mortem* analysis phase is in charge of interpreting the recorded events. This two phase mechanism permits the library to separate the recording of a function call from its interpretation. Moreover, a post-mortem analysis also reduces the overhead of profiling a program; during the execution of the application, the analysis tool should avoid performing time-consuming tasks such as computing statistics or interpreting function calls. It thus allows the user to interpret a function call event, and so a complete execution trace, in different ways depending on the point he/she wants to focus on, just by using different interpretation modules provided by the EZTRACE.

4.3.1 Tracing the Execution of an Application

During the execution of the application, EZTRACE intercepts calls to the functions specified by *plugins* and records events for each of them. Depending on the type of

```
int submit_jobs(int nb_jobs)
BEGIN
 ADD_VAR("Number of jobs", nb_jobs)
 CALL_FUNC
 EVENT("New jobs")
END

void do_work()
BEGIN
 RECORD_STATE("Working")
END
```

Fig. 4.1 Example of function instrumentation using the script language

functions, EZTRACE uses two different mechanisms for interception. The functions defined in shared libraries can be overriden using LD_PRELOAD. When the EZTRACE library is loaded, it retrieves the addresses of the functions to instrument. When the application calls one of these functions, the version implemented in EZTRACE is called. This function records events and calls the actual function. The LD_PRELOAD mechanism cannot be used for functions defined in the application since there is no symbol resolution. In that case, EZTRACE uses the DYNINST [2] tool for instrumenting the program on the fly. Using DYNINST, EZTRACE modifies the program to record events at the beginning and/or at the end of each function to instrument.

For recording events, EZTRACE relies on the FXT library [5]. FXT provides efficient support for recording traces of actions both in kernel and user space. However, EZTRACE only uses the user space feature. In order to keep the trace size as compact as possible, FXT records events in a binary format that contains only the minimum amount of information: a timestamp, an *event code* and optional parameters. During the initialization, FXT pre-allocates a buffer. When recording an event, FXT uses atomic operations for ensuring the trace consistency when multiple threads are used. At the end of the application, FXT flushes the trace to the disk.

4.3.2 Instrumenting an Application

Since EZTRACE uses a two-phases mechanism, plugins are organized in two parts: the description of the functions to instrument, and the interpretation of each function call. During the execution of the application, the first part of the plugin is in charge of recording calls to a set of functions as described in Sect. 4.3.1. The second part of the plugin is in charge of adding semantic to the trace. EZTRACE provides plugins for major parallel programming libraries (MPI, OpenMP, PThread, etc) but also allows user-defined plugins designed for custom libraries or applications. For example, the PLASMA linear algebra library [1] is shipped with an EZTRACE plugin.

In order to ease the creation of a plugin, we designed a compiler that generates EZTRACE modules from a simple script file. As depicted in Fig. 4.1, such a script consists of a list of functions to instrument and the interpretation of each function. In this example, when the function `submit_jobs` is called, EZTRACE increases the value of a counter, calls the original function, and creates an event. A call to `do_work` is represented as a change of the state in the output trace. While some performance analysis tools allow users to specify a set of custom functions to instrument, choosing the representation of the corresponding function calls is usually impossible. Our script language gives the possibility to the users to easily create new EZTRACE modules. Since the compiler generates C files, advanced users can tune the created module to fit their needs.

4.4 Creating Trace Files with GTG

During the *post mortem* analysis phase, EZTRACE browses the recorded events and interprets them. It can then generate statistics – such as the length of messages, the duration of critical sections, etc. – or create a trace file for visualizing the application behavior. For generating trace files, the post-mortem analysis of EZTRACE relies on the Generic Trace Generator (GTG) library.[1] GTG provides an high-level interface to generate traces in different format such as Paje or OTF. This permits EZTRACE to use a single interface for creating traces in multiple formats. Thus, it can generate PAJ traces or OTF files without any modification, just giving at runtime the desired format. This high-level interface can also be used for any application supervision by any user who cares less about tracing performance such as to follow memory consumption and want to combine the tracing and the post-mortem conversion steps.

4.4.1 Overview of GTG

Although trace formats are different, most of them rely on the same structures and provide similar functionalities as it is depicted in Fig. 4.2. A set of hierarchical *containers* (1) represents processing entities such as processes, threads, or GPUs. These containers have *states* (2) that depict events that start at time T_1 and end at time T_2 – the execution of a function, the processing of a computing kernel, a pending communication, etc. – Some *events* (3) (sometimes defined as *markers*) are immediate (i. e. $T_1 = T_2$), and can represent the release of a mutex, the submission of a job, etc. Most trace formats also provide a way to track a *counter* (4) such as the total allocated memory, the number of pending jobs or the number of floating point operations per second. In order to symbolize the interaction between containers, trace formats often provide a *link* (5) feature: a couple of *events* that may happen

[1] Available under the CeCILL-C license at http://gtg.gforge.inria.fr/

Fig. 4.2 Features commonly provided by trace formats

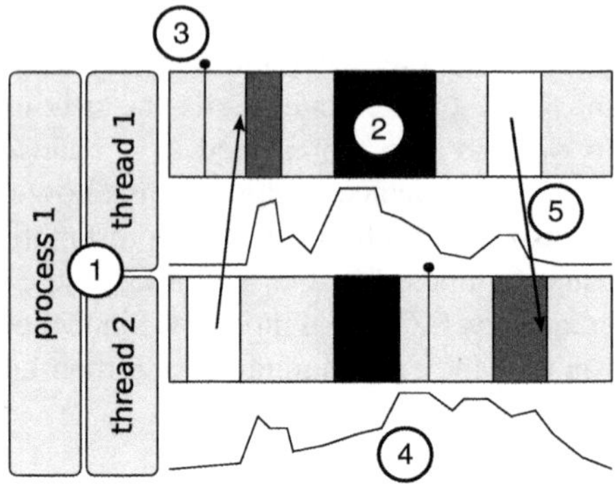

on different *containers*. This permits the viewer tool to represent for example: communications between processes, or signals between threads.

GTG provides a simple interface for manipulating these features. This interface then calls one of the available modules depending on the output trace format.

4.4.2 Interaction Between GTG *and* EZTRACE

Once EZTRACE is running along with the application, FXT traces are generated. The second part of EZTRACE is based on GTG, and transforms the raw traces to real meaningful traces. First, a meaning is added (for example 42 represents an `MPI_Send` request according to the MPI plugin). The semantic can represent links, events, states, etc. The hierarchical structure of the API functions is inspired by the generic PAJ trace format, thus any trace format should be matched. For instance, OTF traces can be generated. The containers can have states (*'This thread is in this function'*), notify events, or count relevant data (number of messages, memory used, number of jobs, etc). This step is based on the plugins (plugins give different meaning to the symbols). Using the EZTRACE convert tool based on GTG, one can add meaning, define containers, and describe what is happening in a function.

4.5 Analyzing Trace Files with VITE

The trace files generated by tools such as GTG or VAMPIRTRACE can be parsed for extracting statistics – such as the average message size – however, understanding the behavior and the performance of an application requires a more convenient tool. In this Section, we present VITE[2] – which stands for *Visual Trace Explorer* – an open-source multi-format trace visualizer.

[2]Freely available at http://vite.gforge.inria.fr/

Fig. 4.3 Modules architecture in VITE

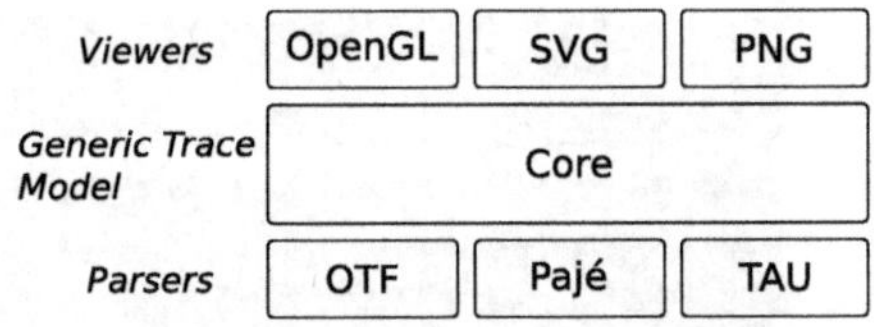

4.5.1 A Generic Trace Visualizer

Originally, the PAJ [7] trace visualizer was designed to analyze parallel applications using a simple yet generic trace format. The decline of PAJ led students to design a new PAJ trace viewer. VITE was designed as a generic trace visualizer, and additional trace formats such as OTF and TAU were added later. To manage multiple formats, VITE relies on a module architecture as depicted in Fig. 4.3.

A set of modules are in charge of parsing traces and filling the generic data structure. VITE implements parsers for several trace formats: OTF, PAJ, extended PAJ (a multiple files PAJ format), and TAU formats. Filling the generic data structure is a critical part of VITE : traces may have millions of events and their processing – storing events, browsing through the event list, finding associated data, etc. – has to be efficient. The last modules are in charge of rendering traces. Such a module uses the data structure to display the trace as requested by the user. A graphical interface based on QT and OpenGL allows an user-friendly browsing of the trace. Additional rendering modules generate SVG or PNG files depicting traces to easily export the results.

Although trace formats are different in their design, most of them provide similar functionalities. VITE implements a generic data structure and manipulates abstract objects representing the different features defined by trace formats. This abstraction permits the developers to easily implement additional parsers for new trace formats, while rendering traces in a homogeneous way.

4.5.2 Displaying Millions of Events

As depicted in Fig. 4.4, VITE is able to display millions of items. To manage such performances, an efficient data structure and an efficient rendering is needed. The data structure is based on binary trees, as depicted in Fig. 4.5. Each tree is built over a sorted list of known size. Hence the construction of the data structure is not optimal (building lists then trees) but the conversion is not so slow (knowing the size of the list, the tree is build in linear time) and it offers an efficient data structure. Moreover, this transformation allows an internal modification, where for instance states are not recorded but state changes are. It avoids storing both the start and the duration of the state: we only store a state change that occurs at time t. This avoids storing an unneeded floating value for each state. Thanks to these trees, any element can

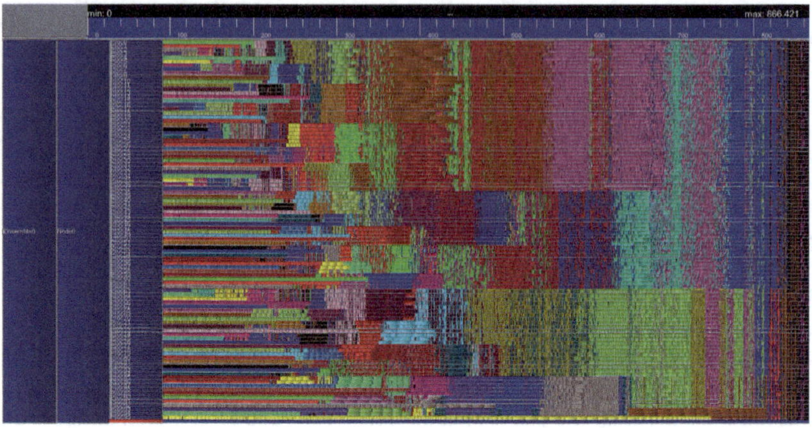

Fig. 4.4 Example of trace display with VITE. The trace contains ten millions events

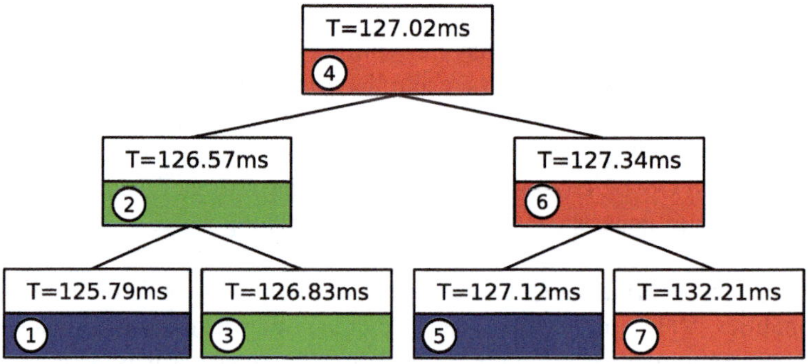

Fig. 4.5 Example of sorted binary tree representation of events

be accessed in logarithmic time and unneeded branches of the tree can be avoided while browsing.

The binary tree structure is also useful for rendering the trace. In order to avoid creating millions of graphical elements, portions of the trace have to be summarized. VITE uses a resolution parameter for eliminating the events that are too small to be rendered: if a node and his father are too close, then the resolution will not be enough for displaying them, and it is useless to keep on browsing all the nodes between the two (the subtree of the node on the same side as the father).

For example, when rendering the binary tree depicted in Fig. 4.5 with a resolution of 1 ms, VITE browses event *# 4*. It then handles *# 2*. Since the interval between events *# 2* and *# 4* is lower than the resolution ($T_{\#4} - T_{\#2} < 1$ ms), event *# 3* is not taken into account. Event *# 1* is then handled normally. Then, VITE processes event *# 6*. Event *# 5* is skipped since it is beneath the resolution ($T_{\#6} - T_{\#4} < 1$ ms) and event *# 7* is handled normally. As a result, the number of elements to display, as well as the number of nodes to browse, is limited increasing the rendering performance. If the user zooms in, the resolution decreases and the same algorithm is used.

The rendering is also critical; OpenGL has been chosen after benchmarking several solutions based on Qt, GTK, SDL, GNUStep and JAVA. Despite the fact that Qt and GTK could provide a better and easier interaction with the trace, the OpenGL engine, with our own mouse placement detection, appeared to be the most scalable solution. Moreover, on some machines, OpenGL can benefit from hardware optimization with the GPU.

4.6 Evaluation

When analyzing the performance of parallel applications that generate millions of events, the performance of the analysis tool is important. The overhead of the instrumentation should be as low as possible, and the visualization tool should allow a smooth browsing of the resulting trace. In this Section, we assess the performance of EZTRACE. We evaluate the raw performance of the instrumentation mechanisms used in EZTRACE on a synthetic benchmark as well as on application kernels.

The results of this evaluation were obtained on the CLUSTER0 platform. It is composed of 32 nodes, each being equipped with two 2.2 GHz dual-core OPTERON (2214HE) CPUs featuring 4 GB of memory. The nodes are running Linux 2.6.32 and are interconnected through MYRINET MYRI-10G NICs. We compare EZTRACE with VAMPIRTRACE in its 5.9 version.

4.6.1 Overhead of Trace Collection

In order to evaluate the raw overhead of program instrumentation, we use an MPI ping pong program. We measure the latency obtained for 16-bytes messages. We instrument this program using the automatic (i.e. using LD_PRELOAD) and manual (i.e. using DYNINST) mechanisms described in Sect. 4.3.1, then we compare the overhead of using EZTRACE or VAMPIRTRACE to the performance obtained without instrumentation. For VAMPIRTRACE, the automatic instrumentation is obtained by using its MPI module. The manual instrumentation is obtained by inserting call to VT_USER_START and VT_USER_END in the application.

Table 4.1 shows the results we obtained. Using VAMPIRTRACE automatic instrumentation degrades the latency by 1.1 μs while the manual instrumentation causes an overhead of 700 ns. The difference is due to the fact that VAMPIRTRACE generates events at the entry and the exit of functions in both instrumentations, but it also generates a *SendMessage* or *ReceiveMessage* event when the MPI module is selected.

Table 4.1 Results of the 16-bytes latency test

Method	Open MPI (μs)	VampirTrace (μs)	EZTrace (μs)
Automatic	4.99	6.12	5.68
Manual	4.99	5.71	5.67

Table 4.2 NAS Parallel Benchmark performance for Class A and B

				VampirTrace		EZTrace		
Kernel	Class	# Proc.	OpenMPI (s)	Time (s)	Overhead (%)	Time (s)	Overhead (%)	# Events/s
BT	A	4	70.57	70.58	0.01	70.39	−0.26	825
CG	A	4	2.64	2.68	1.52	2.68	1.52	12,546
EP	A	4	9.61	9.69	0.83	9.72	1.14	5
FT	A	4	6.63	6.67	0.55	6.62	−0.20	22
IS	A	4	0.63	0.64	2.13	0.62	−1.06	482
LU	A	4	42.08	42.15	0.17	41.39	−1.64	12,282
MG	A	4	5.04	5.06	0.46	5.07	0.66	2,978
SP	A	4	166.25	165.94	−0.18	166.32	0.04	696
BT	B	36	26.08	25.83	−0.97	26.37	1.10	59,350
CG	B	32	16.29	16.46	1.02	16.60	1.88	192,667
EP	B	32	4.81	4.79	−0.42	4.76	−1.04	81
FT	B	32	11.76	11.61	−1.30	11.55	−1.81	255
IS	B	32	0.97	0.96	−1.03	0.96	−1.03	2,580
LU	B	32	33.75	34.11	1.07	33.67	−0.24	–
MG	B	32	2.14	2.16	0.78	2.13	−0.62	215,515
SP	B	36	51.18	51.98	1.57	52.07	1.75	59 922

Instrumenting the application with EZTrace causes an overhead of 700 ns for both mechanisms. This is because EZTrace records events at the entry and the exit of functions for both manual and automatic modes. The *SendMessage* and *ReceiveMessage* events are generated during the *post mortem* phase.

4.6.2 NAS Parallel Benchmarks

In order to evaluate the overhead of EZTrace on more realistic computing kernels, we also measure its performance for NAS application kernels. The experiment was carried out with 4 computing processes for Class A and 32 processes (or 36 for BT and SP that require a square number of processes) for Class B. We instrument MPI functions of these kernels with EZTrace and VampirTrace automatic modules.

Table 4.2 summarizes the results we obtained. Since EZTrace *post mortem* phase crashes for the LU kernel for Class B, the number of events in the resulting OTF trace is not reported. The results show that instrumenting these kernels with EZTrace or VampirTrace does not significantly affect the performance:

variation of the execution time is less than 2 %. This experiments also show that intensive event recording kernels – such as MG or CG for Class B – do not suffer from the overhead of the instrumentation.

4.7 Conclusion and Future Work

Programming a parallel application that efficiently exploits a supercomputer becomes more and more tedious due to the increasing complexity of hardware – multicore processors, NUMA architectures, GPGPUs, etc. – and the use of hybrid programming models that mix MPI, OpenMP or CUDA. Tuning such an application requires the programmer to precisely understand its behavior.

We proposed in this paper an open-source tool-chain for analyzing the performance of modern parallel applications. This software suite is composed of EZTRACE – a generic framework for instrumenting applications –, GTG – a tool for generating traces in multiple formats –, and VITE – a trace visualizer that supports several trace formats –. These tools were designed to provide an open-source alternative to other performance analysis tools, while allowing interoperability with other tools such as Vampir or TAU. The evaluation shows that this genericity does not imply extra overheads since EZTRACE provides similar performance when compared to VAMPIRTRACE.

In the future, we plan to study more precisely the performance of the whole software suite and to improve it. Additional modules are to be developped in EZTRACE in order to allow the analysis of programs running CUDA or OpenCL. We also plan to improve EZTRACE performance analysis capabilities so that it can detect programming or runtime issues such as network congestion or insufficient overlap of communication and computation. Future work concerning GTG includes the support for other trace formats – such as TAU – and enhancing the API. We also plan to merge VITE and TRIVA [12] projects. TRIVA is based on PAJ software and provides new ways of displaying information such as treemaps, or network graphs that will benefit to ViTE. On the other side, TRIVA will benefit from the multi-format parser and from the OpenGL display.

References

1. Agullo, E., Demmel, J., Dongarra, J., Hadri, B., Kurzak, J., Langou, J., Ltaief, H., Luszczek, P., Tomov, S.: Numerical linear algebra on emerging architectures: the plasma and magma projects. In: Journal of Physics: Conference Series, vol. 180, p. 012037. IOP Publishing, Bristol (2009)
2. Buck, B., Hollingsworth, J.: An API for runtime code patching. Int. J. High Perform. Comput. Appl. **14**(4), 317–329 (2000)
3. Caubet, J., Gimenez, J., Labarta, J., DeRose, L., Vetter, J.: A dynamic tracing mechanism for performance analysis of OpenMP applications. In: Eigenmann, R., Voss, M.J. (eds.) OpenMP Shared Memory Parallel Programming, pp. 53–67. Springer, Berlin/London (2001)

4. Chan, A., Gropp, W., Lusk, E.: An efficient format for nearly constant-time access to arbitrary time intervals in large trace files. Sci. Program. **16**(2–3), 155–165 (2008)
5. Danjean, V., Namyst, R., Wacrenier, P.: An efficient multi-level trace toolkit for multi-threaded applications. In: Euro-Par 2005 Parallel Processing, pp. 166–175. Springer, Berlin/New York (2005)
6. Decugis, S., Reix, T.: NPTL stabilization project. In: Linux Symposium, vol. 2, p. 111. Ottawa, Canada (2005)
7. de Kergommeaux, J., de Oliveira Stein, B.: Pajé: an extensible environment for visualizing multi-threaded programs executions. In: Euro-Par 2000 Parallel Processing, pp. 133–140. Springer, Berlin (2000)
8. Geimer, M., Wolf, F., Wylie, B., Ábrahám, E., Becker, D., Mohr, B.: The scalasca performance toolset architecture. Concurr. Comput. Pract. Exp. **22**(6), 702–719 (2010)
9. Knüpfer, A., Brendel, R., Brunst, H., Mix, H., Nagel, W.: Introducing the open trace format (OTF). In: Computational Science–ICCS 2006, pp. 526–533. Springer, Berlin/New York/ Heidelberg (2006)
10. Knüpfer, A., Brunst, H., Doleschal, J., Jurenz, M., Lieber, M., Mickler, H., Müller, M., Nagel, W.: The Vampir performance analysis tool-set. In: Tools for High Performance Computing, pp. 139–155. Springer, Berlin (2008)
11. Muller, M., Knüpfer, A., Jurenz, M., Lieber, M., Brunst, H., Mix, H., Nagel, W.: Developing scalable applications with Vampir, VampirServer and VampirTrace. In: Proceedings of the Minisymposium on Scalability and Usability of HPC Programming Tools at PARCO, Jülich, Germany. 4 to 7 september (2007)
12. Schnorr, L.M., Huard, G., Navaux, P.O.: Triva: interactive 3d visualization for performance analysis of parallel applications. Future Gener. Comput. Syst. **26**(3), 348–358 (2010)
13. Shende, S., Malony, A.: The TAU parallel performance system. Int. J. High Perform. Comput. Appl. **20**(2), 287 (2006)
14. Trahay, F., Rue, F., Faverge, M., Ishikawa, Y., Namyst, R., Dongarra, J.: EZTrace: a generic framework for performance analysis. In: IEEE/ACM International Symposium on Cluster, Cloud and Grid Computing (CCGrid), Newport Beach. IEEE, Piscataway (2011)

Chapter 5
Debugging CUDA Accelerated Parallel Applications with TotalView

Chris Gottbrath and Royd Lüdtke

Abstract CUDA introduces developers to a number of concepts (such as kernels, streams, warps and explicitly multi-level memory) beyond what they are used to in serial, parallel and multi-threaded applications. Visibility into these elements is critical for troubleshooting and tuning applications that make use of CUDA. This paper will highlight CUDA concepts implemented in CUDA 3.0–4.0, the complications they introduce for troubleshooting, and how TotalView helps the user deal with these new CUDA specific constructs.

5.1 Introduction

NVIDIA's CUDA language extension provides a very interesting opportunity for scientists and developers to accelerate the runtime performance of computationally intensive applications running on a single workstation or on a cluster of dedicated nodes augmented with GPU cards. However, as developers begin adapting their applications for the GPU, they are finding out that harnessing the available performance requires mastering the concepts of data parallel programming and developing a fairly sophisticated understanding of the GPU architecture. Parallel development and debugging tools are being adapted to help developers tackle both of these objectives. This paper talks about work that has been done to enhance TotalView so that users can seamlessly debug apps that leverage CUDA in the context of multiple tiers of parallelism: device host, and cluster.

C. Gottbrath (✉) · R. Lüdtke
Rogue Wave Software, Boulder, CO, USA
e-mail: chris.gottbrath@roguewave.com

H. Brunst et al. (eds.), *Tools for High Performance Computing 2011*,
DOI 10.1007/978-3-642-31476-6_5, © Springer-Verlag Berlin Heidelberg 2012

5.1.1 HPC Debugging Challenges

HPC applications, which typically run as distributed parallel applications on cluster type machines, require significant effort to write. Scientists and developers need to think about breaking tasks down into small units that can be distributed across that cluster. They need to pay special attention to the sequence of operations and the data and resource dependencies that different parts of the application have. They need to be conscious not just about what data is needed where, but also about the bandwidth and latency involved with moving data around the cluster.

Frameworks, libraries and languages like MPI, UPC, Global Arrays, Co-Array FORTRAN, OpenMP, and pthreads provide powerful and flexible mechanisms that can be used to construct parallel applications and address the challenges of instantiating and managing parallel programs, providing synchronization and data movement between parallel tasks as required. Regardless of whether it is implicit or explicit, there is a tremendous degree of complexity that arises from parceling up the program and data while ensuring data movement and appropriate (but not too much) synchronization. This complexity becomes an issue when it comes to validating, troubleshooting and debugging applications.

Debugging has always been challenging; there is an old adage that debugging a bit of code is at least twice as hard as writing that same bit of code and that statement was primarily made in the context of serial applications running on a single core. Taking a given algorithm and making it parallel frequently means introducing (1) additional code and (2) more complicated data structures to facilitate distributed execution and data movement. Both cases provide additional opportunities for errors to creep in.

Parallel applications execute many different bits of code at the same time across a large set of distinct nodes. They frequently perform computations that are interconnected enough that an error or oversight anywhere may either cause a local failure or set in motion a series of events that causes incorrect data to be transmitted to another node, leading to a wrong answer or crash somewhere else in this distributed system. A failure to ensure proper synchronization can lead to failures that may happen only rarely when natural perturbations introduced by the system or other factors external to the program lead to rare and unexpected sequencing of operations. Furthermore, synchronization-dependent errors may become visible only when the program is run at or above some specific scale or is sensitive to details about how parallel tasks are mapped to compute resources.

All of this adds to the complexity of validating, debugging and tuning parallel applications. As if distributed architectures weren't enough, recent trends have introduced heterogeneous computational elements into the mix.

5.1.2 CUDA and Heterogeneous Acceleration Architectures

Recently, as general-purpose processors have grown more complicated and physical limits having to do with thermal dissipation have served to limit raw clock speed increases, there has been a growing interest in exploring heterogeneous architectures. Such architectures attempt to blend the benefits of general-purpose cores that excel at executing a wide range of computation with a set of cores specifically designed to execute more limited types of computations. These special-purpose cores typically focus on performing integer and floating point calculations on vectors of data and may trade off some of the sophisticated predictive control flow capabilities that are featured on the general purpose cores. A few years ago IBM and Toshiba introduced the Cell processor, which blended a PowerPC-based general purpose computing element in a processor package with a set of special vector processors called Synergistic Processing Elements (SPEs). NVIDIA provides a solution that uses their extremely capable, mass-market Graphics Processing Units (GPUs) in the vector processor in conjunction with traditional mass-market Intel or AMD CPUs.

There are several different techniques for taking advantage of GPUs as computational accelerators. In the 1990s curious users who found clever ways to repurpose the OpenGL image processing language to accomplish general-purpose computations. Later NVIDIA introduced the proprietary CUDA for C and C++ language extensions. AMD and Apple sponsored work resulting in OpenCL, and independent companies like PGI and CAPS introduced techniques like the PGI-accelerated Fortran and CAPS HMPP. CUDA has a unique position in the market in terms of widespread adoption and a mature, though still rapidly evolving, tool chain and runtime environment. While the jury is still very much out on which approach will be the favored one several years down the road, CUDA is how most ISVs and users are implementing GPU acceleration for NVIDIA hardware today.

5.1.3 Challenges Introduced by CUDA

Programming for CUDA introduces a series of unique challenges that HCP developers need to overcome to successfully harness the power of the GPU and reduce the time that it takes their applications to run and generate results.

The first challenge is a lack of abstraction; CUDA generally exposes quite a bit about the way that the hardware works to the programmer. In this regard, it is often compared to assembly language for parallel programming. Explicit concepts such as thread arrays, discussed below, map directly to the way the hardware groups computational units and local memory. CUDA compilers and runtimes prior to 3.2 don't support thread stacks, so programmers had to avoid recursion and generally be aware that any functions they wrote in CUDA device code would be inlined for execution. The memory model requires explicit data movement between the host processor and the device and CUDA makes explicit use of a hierarchy of memory address spaces, each of which obeys different sharing rules. This is more low-level detail than the typical application or scientific programmer has to deal with when programming in C, C++ or Fortran and also raises significant concerns with regard to portability and maintainability of code.

The second such challenge is that the programming model for CUDA is one of data parallelism rather than task parallelism. When divvying up work across the nodes of a cluster, HPC programmers are used to looking for and exploiting parallelism at a certain level. The amount of data to assign to each node depends on a number of factors including computational speed of the node, the available memory, the bandwidth and latency of the interconnect and the frequency with which results need to be shared with other processes. Since processors are fast and network bandwidth is relatively scarce, the balance is typically to put quite a bit of data on each compute node and to move data as infrequently as possible. CUDA invites the programmer to think about parallelism of a completely different order, encouraging the developer to break the problem apart into units that are much smaller, often as small as the computation that might be required to assign a single variable into an array. This is often many orders of magnitude more parallelism than was previously expressed in the code and requires reasoning somewhat differently about the computations themselves.

The third challenge lies in dealing with data movement and multiple address spaces. NVIDIA GPUs are external accelerators attached to the host system via the PCI bus; each GPU has both its own onboard memory and also features smaller bits of memory attached to each one of the compute elements (see the architecture review below for more detail). While there are now mechanisms to address regions of host memory from device kernels, such access is very slow compared to accessing device memory. CUDA programs therefore use a model in which code running on the host processor prepares and explicitly dispatches work to the GPU, pauses for the GPU to complete that work, then reads the resulting data back from the device. Both the units of code representing computational kernels and the associated data on which those computational kernels will execute are dispatched to the device. The data is moved, in whole or part, to the device over the PCI bus for execution and as results are produced they need to be moved, just as explicitly, back from the device to the host computer's main memory.

The device has different kinds of instance memory: either local and reserved for specific sets of computing elements, or globally addressable from any of the computing elements on the CUDA device. CUDA makes this distinction explicit, and requires that developers designate the location of variables at variable declaration time. A variables location in this memory hierarchy determines how it can be used, forcing the developer to be cognizant of these issues.

5.1.4 The TotalView Debugger

TotalView is a highly interactive graphical source code debugger that provides developers working in C, C++ or Fortran with a way to explore and control their programs. Originally designed to debug one of the first distributed memory architectures, the BBN Butterfly, TotalView has been continually enhanced with a focus on multi-process and multithreaded applications. Today TotalView is used to develop, troubleshoot and maintain applications in a wide range of situations. One group uses it to develop Linux-based commercial embedded computing applications consisting of a single process with a 100 or more separate threads that simultaneously interact with a graphical user interface (GUI), sensors, online databases and network services. Other users troubleshoot sophisticated mathematical models of complex physical systems in astrophysics, geophysics, climate modeling and other areas. These models typically consist of thousands of separate processes that run on large-scale, high performance computing (HPC) clusters on the top 500 list. In both these use cases, TotalView provides the users with a debugging session in which they can examine in detail any one of the many threads or processes that make up the application, controlling each thread or process singly or as a part of various predefined and user-defined groups, and displaying multiple types of data and state information from across many processes and threads. With runtimeperformance – conscious developers now exploring CUDA as a way to accelerate computation, Rogue Wave Software has enhanced TotalView to support users who are developing and debugging CUDA C/C++ code.

5.2 TotalView for CUDA

This section of the paper details what Rogue Wave Software has done to adapt the TotalView debugger to support CUDA development and debugging.

5.2.1 Previous Experience: Cell

The IBM Cell processor technology was a precursor in several ways to GPU technology. Rogue Wave Software worked with Los Alamos National Lab (LANL) to provide support for their leadership-class Roadrunner system. The Cell processor embraces heterogeneity, providing a general purpose POWER core and a set of eight synergistic processing elements that are somewhat simpler and more restricted than the POWER core but each capable of performing computations within a separate memory environment. As with the GPU, user programs run on the general purpose core, which calls on the special purpose processors to accelerate certain types of computation.

The architecture of the Cell presents two general challenges for developers writing new software for or porting existing software to the Cell. The first challenge is in breaking the key components of a program into small chunks that can execute on the eight Synergistic Processing Elements (SPEs). In numerical programs, this often means dividing the data into small independent units. In other cases, individual tasks or pipeline stages may be delegated to specific SPEs. This issue of problem decomposition is similar to that of adapting an application to a distributed memory cluster environment, although the granularity is different due to the limited memory space directly available to each SPE.

The second challenge has to do with moving data between different parts of the processor. Each of the eight SPE cores that is part of the Cell processor has its own independent registers and a small amount of local memory (256 KB) used for storing instructions and data. This memory acts similarly to a cache in a general-purpose processor, in that it has a limited size and can be read and written very quickly. Unlike a cache, however, its contents are directly managed by the application. The code units running on the SPE elements are allowed to initiate direct memory access operations that can copy chunks of data from the main memory into the SPE local store or back to main memory from the local store. The same memory is used for the machine instructions and global memory, heap memory and stack. The heap and the stack change size over time and because there is no memory protection, the programmer must take care to avoid collisions between these different memory ranges. Because each processor has a completely separate local store, the structure of a program for the Cell is very different than the structure of a traditional multi-threaded application, where all the threads share the same memory.

In order to support the Cell processor TotalView extended the process and thread model, allowing it to accommodate the idea that a process might contain a heterogeneous mixture of threads, some of them running in the shared address space of the host processor and some of them running off in the separate and more constrained space of the SPE cores. That had implications on the group models that TotalView uses to decide where to plant breakpoints and what processes to hold and what processes to run when performing process control operations. The result was a very natural and powerful user interface paradigm where the user could quite easily switch between host processor threads and the accelerator threads running on the SPEs.

5.2.2 The NVIDIA GPU Architecture and CUDA

The CUDA hardware model has some significant similarities and differences from the Cell processor. Where the Cell had eight synergistic processing elements, the NVIDIA GPU has an array of streaming multiprocessors (SMs) each of which provides an execution environment with a bit of local memory and a number of streaming processors (SP). Older hardware may have as few as 8 SPs per SM and newer hardware has up to 48. Devices such as the NVIDIA Tesla C2070 have 14 SMs with 32 SPs each and can execute up to 448 operations in parallel.

Each one of those streaming processors may be working on different data and produce different results but they are not all executing independently. The NVIDIA hardware introduces the concept of a warp, which is a set of 32 threads that execute the same instruction together. Branching within the warp is handled by stalling those threads that are not participating in a branch to be resumed when the shared program counter is again executing instructions that are part of the thread occupying that lane's execution path.

Logically threads are grouped together into arrays of up to 512 threads (16 warps) that execute on an SM and share a bit of memory. The geometry of the array is configurable based on the code and an individual thread is identified by a unique combination of three coordinate indices. Since each thread array executes on a single SM, CUDA programs will typically try to start up many more than one thread array at a time. CUDA provides a way to address sets of thread arrays using a second set of three coordinate indices that are referred to as the grid coordinates. Threads are grouped together into arrays. Arrays are grouped together in a grid. The full combination of six dimensional grid and array coordinates serve to fully specify a unit of work within a discrete CUDA computation.

5.2.3 The TotalView Model: Extended for CUDA

This section of the paper details Rogue Wave Software's adaptations to the TotalView debugger in support of CUDA development and debugging.

TotalView has been extended to provide CUDA support in the same manner that it was extended for Cell. The new version of TotalView builds on and extends the model of processes made up of threads to incorporate CUDA. There are two major differences: first, CUDA threads are heterogeneous, and second, TotalView shows a representative CUDA thread from among all those created as part of running a routine on the GPU.

The GPU thread is shown as part of the process alongside other pthreads or OpenMP threads. Threads created by OpenMP or pthreads in any single process share an executable image, a single memory address space and other process level resources like file and network handles. This special CUDA thread, on the other hand, executes a different image, on a set of processors with a different instruction set, in a completely distinct memory address space.

Because of the large number of CUDA threads and the fact that only a small number of this large set of logical threads are likely to actually be instantiated on the hardware at any given moment in time, the UI has been adapted to display the thread specific data from a user-selectable representative CUDA device thread.

CUDA is heavily used in conjunction with distributed multi-process programming paradigms such as MPI. As such, the CUDA support in TotalView is complimentary and additive with the MPI support in TotalView. Users debugging an MPI+CUDA job see the same view of all their MPI processes running across the cluster and when they focus on any individual process they can see both what is happening on the host processor and what is happening in the CUDA kernels running on the GPU.

5.2.4 Challenges and Features

This section will discuss some of the challenges encountered as TotalView was adapted to allow it to debug CUDA accelerated applications and will highlight some of the features that were introduced as a result.

5.2.4.1 Extending the Thread Model

TotalView already has a very powerful model for multi-process and multithread applications. Each program is modeled as a set of processes, which may run on one or more hosts and which each are comprised of one or more threads. TotalView has features that allow the user to examine and control each thread separately. Most programs are comprised of processes that have anywhere from one to a handful of threads. A few ambitious programs may have a 100 or more threads. The CUDA runtime allows each thread of a UNIX process to dispatch work to be done on the attached GPU device. This work is encapsulated in units of work referred to abstractly as "kernels" and the parallel architecture of the GPU allows it to execute hundreds of instances of these kernels, called device threads, at a time. In order to hide latencies the programmer is encouraged to request the creation of large numbers (many thousands) of such device threads.

The device cannot actually process all of these threads as the same time; instead it "streams" them – scheduling them onto the hardware in batches, running each batch to completion while loading the preconditions for the next batch so that it can immediately run. This streaming technique hides much of the cost of data movement and context creation.

Ideally each CUDA kernel thread is performing self-contained unit of work with clearly defined inputs and with no side effects or dependencies on anything that is meant to happen within any of the other kernel threads that are being executed at the same time. That allows the device and the device runtime environment a large degree of freedom in scheduling the kernels. If resources allowed, the runtime could create all the device threads ahead of time and run them all simultaneously. Alternately it could create one thread at a time, running it to completion and then eliminating it (leaving just its output data). Generally the model allows the runtime to create and run device threads in any order or grouping that make sense based on hardware resources and the data being processed. This device kernel thread concept is very different from host processor threads. The creation of a thread in pthreads or OpenMP is a significant and heavyweight operation; best programming and thread runtime practices involve keeping those threads around for a relatively long period of time and using various techniques to have them operate on data structures that are shared across and between threads. The fact that multiple threads access the same data structures is the key feature and key challenge to multithreaded programming with pthreads or OpenMP.

While CUDA kernel threads are encouraged to be self-contained they are not forced to be. They can access (both reading from and writing to) global and shared memory. This is important in that it avoids having the CUDA program duplicate certain input data structures, but it means that there is the possibility for race conditions and device threads reading incorrect or invalid data.

In order to give visibility into the dynamically changing set of CUDA kernel threads without overwhelming the user, TotalView creates a single thread object for each kernel invocation, called a GPU focus thread, which is displayed next to the host threads in the debugger's display. The user selects this thread to get a view into one of the device threads that are currently executing the kernel on the GPU device. TotalView gives host threads a positive debugger thread ID and CUDA threads a negative thread ID. The initial host thread in process "1" might be labeled "1.1" and the CUDA thread is labeled "1.-1". See Fig. 5.1 for just such a pair of threads.

At any time the user can control which of the currently active CUDA threads is displayed through the GPU focus thread. This selection is done using a new set of controls, called the GPU thread selector, on the TotalView Process Window. The GPU thread selector reports the device thread currently being displayed, and can be used to change that selection. CUDA threads can be specified using one of two coordinate spaces. One is the logical coordinate space that is in CUDA terms of grid and block indices: <<<(Bx,By,Bz),(Tx,Ty,Tz)>>>. The other is the physical coordinate space that is in hardware terms of device number, streaming multiprocessor (SM) number on the device, warp (WP) number on the SM, and lane (LN) number on the warp. See the GPU thread selector in Fig. 5.1.

Any given thread has both a thread index in this 4D physical coordinate space, and a different thread index in the 6D logical coordinate space. These indices are shown in a series of spin boxes in the Process Window, next to a button that indicates which coordinate system is currently displayed. Pressing this button switches between the two numbering systems, but does not change the actual thread.

It isn't always obvious how the logical and physical coordinates are mapped to one another, nor is it always clear what logical coordinate range designates the set of threads that are currently mapped to hardware. Therefore TotalView added a CUDA device display which brings up a separate window specifically designed to display the hardware's capabilities (in terms of how many streaming multiprocessors it has and their compute capability rating) and the logical threads currently mapped to the hardware resources (Fig. 5.2).

5.2.4.2 Dealing with Different Memory Spaces

TotalView can display variables and data from a CUDA thread. CUDA variables for a device kernel function are displayed in the stack frame on the Process Window in the same manner that variables are displayed in C, C++ or Fortran code on the host processor. The idioms of hovering over variables with the mouse or diving on variables to open a Variable Window both work just like they do on the host thread, as does the TotalView expression list and expression system. One thing to note is

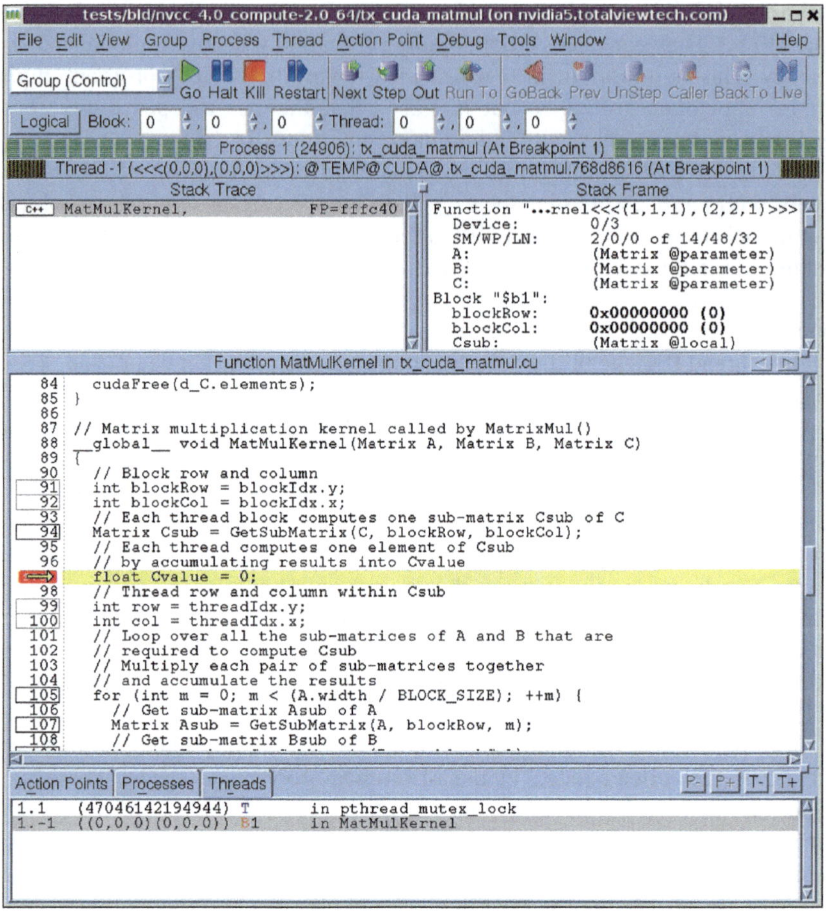

Fig. 5.1 TotalView Process Window focused on a CUDA device thread. Note the CUDA thread selector between the toolbar with the stepping controls and the process status ribbon; the CUDA-specific information in the stack frame, including the qualifiers such as local on the type of variable Csub; and the CUDA device thread 1.-1 and the host thread 1.1 listed in the Threads Pane at the *bottom* of the window

that when you use the CUDA thread selector to change the focus you will see the values of variables in the newly selected thread replace those from the previous thread in all the aforementioned places.

Naturally, all variables have addresses, but depending on the variables type, those addresses may refer to any one of many different places. That's because CUDA uses an explicitly hierarchical memory. Each thread has its own local memory; so if variable X is a local variable and it has address 0×14 each thread can have its own instance of X, all with the same address but each referring to a private instance of that variable with a different value. Similarly there is a bit of memory associated with each cooperative thread array and the streaming multiprocessor that the

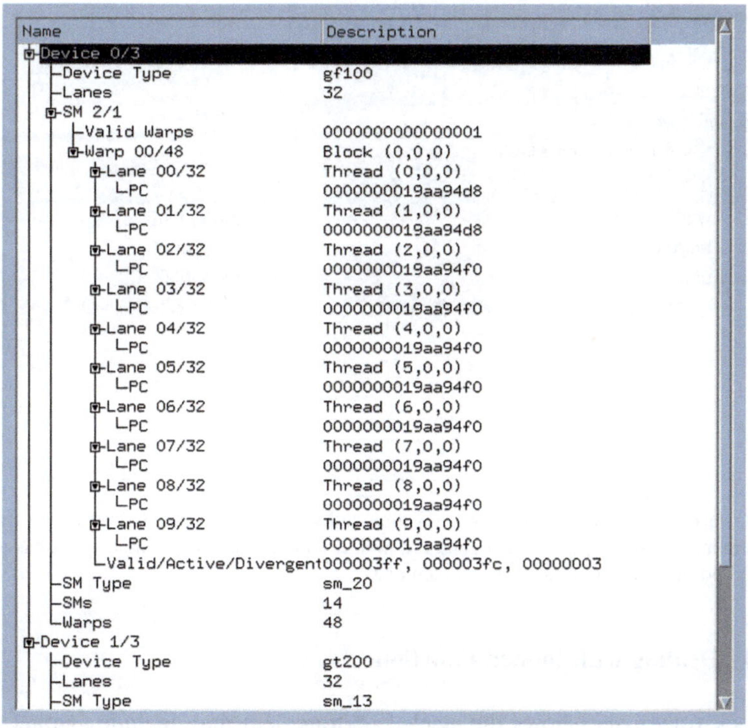

Fig. 5.2 Device Status Window, showing the device hierarchy starting with a numbered set of CUDA devices that are comprised of SMs that include Warps made up of Lanes. This particular display highlights the 1st of three devices on a system, with four active SMs and their component Warps and Lanes. The other key information that is presented here is how these hardware resources are mapped to the logical resources of the CUDA program itself. CTA blocks are mapped to specific SMs and threads are mapped to warps and lanes. In the example shows the thread with device coordinates (Device, SM, Warp, Lane) = (0,2,0,3) is mapped to block (0,1,0) thread (1,1,0)

CTA is running on. Variables that reference into that memory refer to data that is shared by all the threads in the CTA. Finally there are variables that refer into the global memory that is addressable by and shared by all the threads running on the device.

TotalView expresses these different kinds of memory (register or local, shared, constant or global memory) through the notion of a storage qualifier in the type system. When looking at variables in CUDA these qualifiers appear next to familiar types such as "int" or "float" so that a local variable X expressing an integer would have type of "int @local" in the TotalView Variable Window. Like types, these qualifiers are something that the user can modify. Casting a variable from "int @local" to "int @global" won't change the pointer but it will use that pointer to look in a different place (the same address in a different memory address space) and will likely refer to a different value. These type qualifiers can be seen both in Fig. 5.1 above and in Fig. 5.3 below.

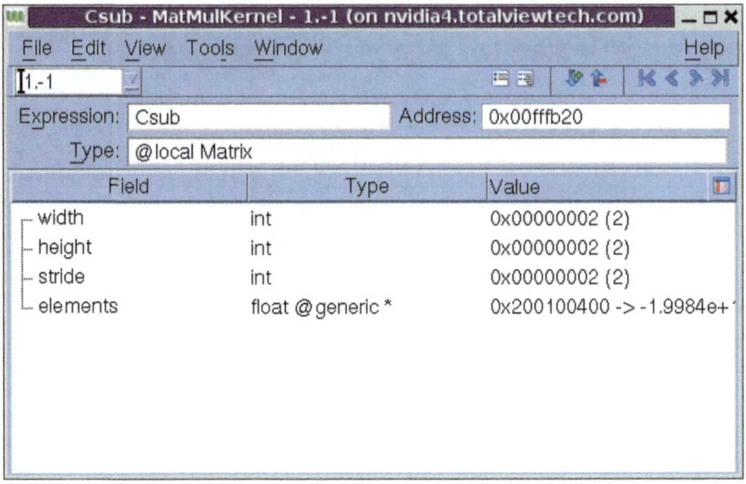

Fig. 5.3 Variable Window displaying a variable called Csub that is at 0x00fffb20 in the thread local memory. It has a type of Matrix and it contains a pointer named elements that has a type of float* and points to an address in global memory

5.2.4.3 Dealing with Inlined Functions

The stack trace pane shows the stack backtrace and may include synthetic stack references for inlined functions. Each stack frame in the stack backtrace represents either the PC location of GPU kernel code, or the expansion of an inlined function. Inlined functions can be nested. The return PC of an inlined function is the address of the first instruction following the inline expansion, which is normally within the function containing the inlined-function expansion.

5.2.4.4 Thread Control and Single Stepping the GPU

TotalView allows you to single-step GPU code just like normal host code, but its important to note that a single-step operation steps the entire warp associated with the GPU focus thread. That's because at the hardware level, all the threads in the warp share a single program counter. It is literally not possible for the GPU to understand the concept of having some of the threads in the warp at location A and the other threads at location B. In some senses this seems like a limitation on the freedom TotalView gives developers to arbitrarily control the execution of host threads. However, the purpose of that control is to allow the developer to explore alternate possible execution scenarios. Since it isn't possible for the threads in the warp to get out of step with one another there is no need to exercise scenarios other than those with synchronized threads across the warp. Different warps, however, might examine or modify global memory and might run with different timing, so it is both valid and sometimes necessary to explore different sequences of execution between distinct warps.

5.2.4.5 CUDA Memory Exceptions

The NVIDIA hardware lacks some of the memory protection logic that developers are likely accustomed to when writing code for the host processor, so mistakes like de-referencing an uninitialized pointer in CUDA code won't result in a segmentation violation, and the program will proceed with undefined consequences. However, NVIDIA provides a mechanism for emulating the memory protection in software; that mechanism can be engaged in TotalView by activating the preference "Enable CUDA Memory Checking". With CUDA memory checking activated you can run the device kernel, and when an invalid pointer operation occurs the debugger will stop the thread with an error message as it does when a host thread encounters a segmentation error. Note that CUDA memory checking is not active when single-stepping the CUDA thread.

5.2.4.6 Conclusion

This paper has focused on two main topics. First, it briefly reviewed CUDA and some of the challenges that CUDA introduces for HPC developers looking to harness GPU processing power to accelerate their applications. These challenges center on the conceptual transitions related to moving to (or more often layering in) fine-grained data parallelism from (or alongside) medium-grained task parallelism, but also extend to dealing with many low-level details about the NVIDIA GPU device architecture that are exposed to the CUDA programmer as additional language concepts. These features and challenges, while certainly not insurmountable, definitely complicate the picture; the increased complexity potentially hinders troubleshooting, validation, and ongoing software maintenance.

The second section of the paper focuses on changes that have been made to adapt the TotalView parallel debugger for CUDA. The paper shows how the TotalView process and thread model has been extended to allow users to easily switch back and forth between what is happening on the host processors, where MPI-level communication occurs and the program sets the stage for the data parallel work to happen on the GPU, and what is happening over on the GPU itself where lightweight threads are being created and destroyed at a furious pace, each one in existence only long enough to compute one data element. It also details how the TotalView variable display and type system have been modified to provide developers a clear understanding of how their data interacts with the various distinct memory address spaces, each with different characteristics and visibility.

Chapter 6
Advanced Memory Checking Frameworks for MPI Parallel Applications in Open MPI

Shiqing Fan, Rainer Keller, and Michael Resch

Abstract In this paper, we describe the implementation of memory checking functionality that is based on instrumentation tools. The combination of instrumentation based checking functions and the MPI-implementation offers superior debugging functionalities, for errors that otherwise are not possible to detect with comparable MPI-debugging tools. Our implementation contains three parts: first, a memory callback extension that is implemented on top of the `Valgrind` `Memcheck` tool for advanced memory checking in parallel applications; second, a new instrumentation tool was developed based on the Intel Pin framework, which provides similar functionality as `Memcheck` it can be used in Windows environments that have no access to the Valgrind suite; third, all the checking functionalities are integrated as the so-called `memchecker` framework within Open MPI. This will also allow other memory debuggers that offer a similar API to be integrated. The tight control of the user's memory passed to Open MPI, allows us to detect application errors and to track bugs within Open MPI itself. The extension of the callback mechanism targets communication buffer checks in both pre- and post-communication phases, in order to analyze the usage of the received data, e.g. whether the received data has been overwritten before it is used in an computation or whether the data is never used. We describe our actual checks, classes of errors being found, how memory buffers are being handled internally, show errors actually found in user's code, and the performance implications of our instrumentation.

S. Fan (✉) · R. Keller · M. Resch
High Performance Computing Center Stuttgart (HLRS), Nobelstrasse 19, 70569 Stuttgart, Germany
e-mail: fan@hlrs.de; keller@hlrs.de; resch@hlrs.de

H. Brunst et al. (eds.), *Tools for High Performance Computing 2011*,
DOI 10.1007/978-3-642-31476-6_6,

6.1 Introduction

Parallel programming with the Message Passing Interface MPI [7] as a distributed memory paradigm is an error-prone process. Great effort has been put into parallelizing libraries and applications using MPI. However when it comes to maintaining software, optimizing for new hardware, or even porting codes to other platforms and other MPI implementations, developers will face additional difficulties [2]. They may experience errors due to hard-to-track interleaving dependent bugs, deadlocks due to communication characteristics, and MPI-implementation defined or even hardware dependent behavior. One class of bugs that are hard-to-track are memory errors, specifically in non-blocking and one-sided communication.

In previous work, we introduced a new implementation of the `memchecker` framework within Open MPI to check for memory problems in parallel applications [4, 11]. It integrates the `Valgrind Memcheck` tool in `memchecker` to observe communication buffers during the communication as well as user specified parameters. These memory checks only observed memory before a communication finishes, which can be refered to as a pre-communication check. However, the functionality of the memory checking framework can be more widely extended. New memory checking can also be implemented for observing the buffers after the communication is finished, which we refer to as post-communication checks, for example, the usage of the received buffer. On the other hand, Open MPI fully supports Windows platforms, so these correctness features could be extended to Windows.

In this paper, we introduce an extension of the `Memcheck` tool of the `Valgrind` tool suite [10] and a new memory debugging tool, `MemPin`, which provides similar memory checking features as `Memcheck` for Windows. Both of these have been integrated into Open MPI for pre- and post-communication checks. The communication buffers errors, such as accessing buffers of active non-blocking operations, writing communicated buffer before reading it, or transferring unused data are being checked and reported. This kind of functionalities would otherwise not be detectable within traditional MPI-debuggers based on the PMPI-interface.

The structure of this paper is as follows: Sect. 6.2 shows the basic idea and functionalities of `Memcheck` and Intel PIN; Sect. 6.3 first introduces the extension of the `Memcheck` and the new Pintool, then it gives an overview of the design and implementation of the memory debugging frameworks using our extensions and the Pintool; Sect. 6.4 shows details of the pre- and post-communication checks for parallel applications; in Sect. 6.5, we show the performance implications of these two scenarios; then in Sect. 6.6 we give examples of the errors, that are being detected and have been detected so far; finally; in Sect. 6.7, we make a comparison with other available tools and conclude the paper with an outlook of our future work.

6.2 Overview of Debugging Tools

6.2.1 *Valgrind*

The tool suite `Valgrind` [10] may be employed on static and dynamic binary executables on `x86/x86-64` and `PowerPC32/64` compatible architectures. It operates by intercepting the execution of the application on the binary level and interpreting the instructions. With this instrumentation, `Valgrind` tools then may deduce information and perform checks of different methodologies.

The system core of `Valgrind` provides a synthetic CPU. When the application starts, `Valgrind` will "trap" the real CPU and run the machine code on its synthetic CPU. Meanwhile, the debugging information is read from the executable and its associated libraries. This instrumentation for the `Valgrind` parser uses processor instructions that do not otherwise change the semantics of the application. By this special instruction preamble, `Valgrind` detects commands to steer the instrumentation. On x86 architecture, the right rotation instruction `ror` is used to rotate the 32-bit register `edi`, by 3, 13, 29 and 19, aka 64-Bits, leaving the same value in `edi`; the actual command to be executed is then encoded with an register-exchange instruction (`xchgl`) that replaces a register with itself (in this case `ebx`):

```
#define __SPECIAL_INSTRUCTION_PREAMBLE                      \
            "roll $3,  %%edi ; roll $13, %%edi\n\t"  \
            "roll $29, %%edi ; roll $19, %%edi\n\t"  \
            "xchgl %%ebx, %%ebx\n\t"
```

`Memcheck`, a heavyweight memory checker in the `Valgrind` tool suite, is well known for tracking memory definedness down to the bit level, which guarantees that partialy defined bytes are also correctly dealt with. It stores two kinds of shadow memory values, for addressability and definedness. `Memcheck` shadows each byte in memory with information that represents whether the byte has been allocated (so-called A-Bits) and for each bit of the byte, whether it contains a defined value (so-called V-Bits). With this AV-bit pair implementation, `Memcheck` is able to provide bit-precision checks of program errors as they run. It tracks the addressability of every byte of memory and the definedness of every bit of data in registers and memory, so that it can detect accesses to unaddressable memory errors and use of undefined values, such as buffer overruns and faulty accesses to the stack. In total, every byte of memory is shadowed with 9 bits (one A bit plus eight V bits). `Memcheck` also tracks all heap blocks allocated with `malloc()`, `new` and `new[]` to detect bad or repeated frees of heap blocks and memory leaks. Arguments to functions like `strcpy()` and `memcpy()`, are also checked for overlaps.

However, the disadvantage of using `Memcheck` is the slowdown of the running application, which is caused by adding code to check every memory access and every value that is computed. The size of the code is usually increased by a factor of 12, and runs 25–50 times slower than natively.

6.2.2 Intel Pin

Intel Pin [6] is a framework for building robust and powerful software instrumentation tools, such as profiling, performance evaluation, and error detection tools. Intel Pin is capable of analyzing and instrumenting application code, and it is easy-to-use, portable, transparent, and efficient for building instrumentation tools (Pintools) written in C/C++. The Pin framework follows the ATOM [12] model that allows the Pintools to analyze the application at the instruction level without detailed knowledge of the underlying instruction set. The framework API is designed to be platform independent in order to make the Pintools compatible across different architectures. It can provide architecture specific details when necessary. The instrumentation process is transparent as both the application and the Pintool observe the original application code. Pin uses techniques like inlining, register reallocation, and instruction scheduling, in order to run more efficiently. The basic overhead of the Pin framework is approximately 10–20 %, and extra overhead might be caused by the Pintool.

Figure 6.1 shows a basic runtime architecture of Pin on Windows. It consists of three main processes, the launcher process, the server process and the instrumented process. The launcher process creates the other two processes, injects the Pin modules, and instrumented code into the instrumented process, then it waits for the process to terminate. The server process provides services for managing symbol information, injecting Pin, or communicating with the instrumented process via shared memory. The instrumented process includes the Pin Virtual Machine Monitor (VMM), the user defined Pintool library, and a copy of the application executable and libraries. The VMM is the core engine of the entire instrumented process that includes a system call emulator, event and thread dispatchers, and also a (Just In Time) JIT compiler. After Pin takes over the control of the application, the VMM coordinates its execution. The JIT compiler instruments the code and passes it to the dispatcher, which launches the execution. The compiled code is stored in the code cache. The Pin tool contains the instrumentation and analysis routines. It is a plug-in and linked with the Pin library, which allows it to communicate with the Pin VMM.

The Pin JIT compiler recompiles and instruments small chunks of binary code immediately before executing them. The modified instructions are stored in a software code cache. It allows code regions to be generated once and reused for the remainder of program execution, in order to reduce the costs of recompilation. The overhead introduced by the compilation is highly dependent on the application and workload [6].

In this paper, we will describe the implementation of `Memcheck` extension and the integration of a newly developed Pintool as the second `memchecker` component in Open MPI, which may help MPI application and Open MPI developers to track erroneous use of memory, such as reading or writing buffers of active non-blocking receive operations, writing to buffers of active one-sided get operations as well as erroneous use of communicated data, such as writing the communicated buffer before reading.

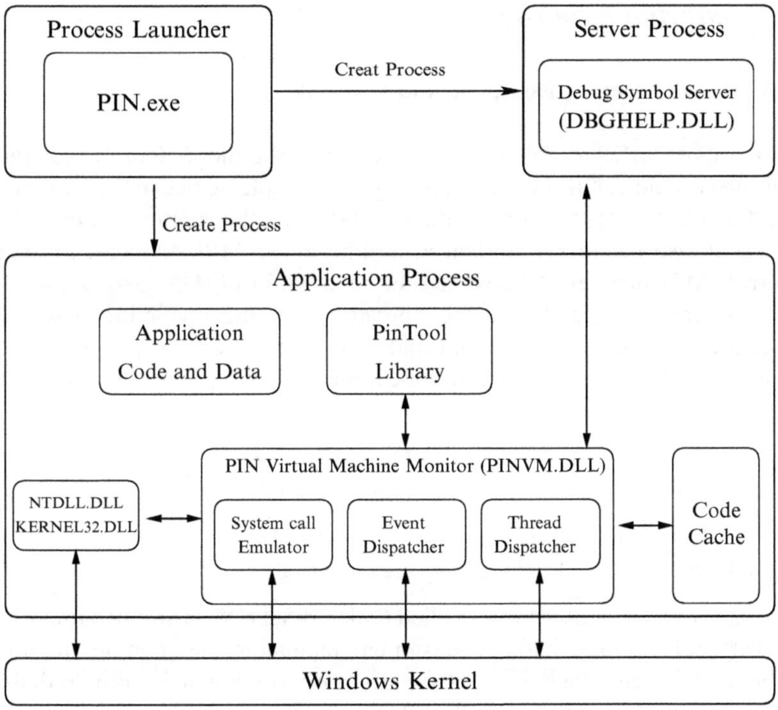

Fig. 6.1 Overview of program execution with an Intel Pin tool on Windows

6.3 Design and Implementation

In previous work, we have taken advantage of an instrumentation API offered by `Memcheck` to find MPI-related hard-to-track bugs in applications (and within Open MPI). In order to allow other kinds of memory-debuggers, such as `bcheck` [1] or Totalview's memory debugging features [14], we have implemented the functionality as a module within Open MPI's Modular Component Architecture [16]. The module is therefore called `memchecker` and may be enabled with the configure-option `--enable-memchecker`.

However, that did not meet all the requirement of advanced MPI semantic memory checking, and the current functionalities of `Valgrind Memcheck` do not suffice. For example, the future MPI standard will change behavior compared to the current version of the standard in that it allows read access to send buffers of non-blocking operations. In order to detect send buffer rewrite errors, `Memcheck` has to know which memory region is readable or writable. On the other hand, checking the communicated buffer is also important, for example, writing the received buffer before reading may cause computation errors. And for performance concerns, data transferred but never used (read) might also be necessary to detect. The extensions for `Memcheck` have been implemented for this purpose.

6.3.1 Valgrind Extensions

6.3.1.1 Make Memory Readable and Writable

Making a user buffer readable or writable will make the memory checking more flexible and accurate, but such functionality is not implemented in the current design of the `Valgrind Memcheck` tool. For MPI parallel memory debugging, this feature will also become important according to the MPI standard, for example, the current MPI standard defines, that the send buffer of `MPI_Send` must remain completely inaccessible during the operation, but the upcoming new standard will release the read restriction and make the buffer readable [9]. Therefore, the extension for `Memcheck` was developed, and two more new `Memcheck` client requests were implemented:

- `VALGRIND_MAKE_MEM_READABLE(addr,len)`
  Make the user specified memory region readable.
- `VALGRIND_MAKE_MEM_WRITABLE(addr,len)`
  Make the user specified memory region writable.

This new extension uses the so-called Ordered Set (OSet) architecture originally from Valgrind's source base. The storage maintains the memory region with readable and writable flags in a binary tree. Overlapping and multiple definitions on the same memory region are handled to avoid redundant records. The basic searching algorithm is a binary search with the starting address of the memory region. When a memory region is being accessed, the memory OSet table will be first searched, if this memory matches any entry in the table, the recorded memory state will be checked so that to eliminate the read or write limit.

6.3.1.2 Callback Extension on Memory Access

In order to perform a precise data transfer evaluation, for example, whether the received data has been correctly accessed and used, it is necessary to have a functionality to monitor on the buffer access for user demand. To achieve this goal, two more new `Memcheck` client requests were implemented:

- `VALGRIND_REG_USER_MEM_WATCH(addr,len,op,cb,data)`
  Registers a callback function for the user specified buffer on specific operation, and the callback function will be called when the operation is triggered.
- `VALGRIND_UNREG_USER_MEM_WATCH(addr,len,cb)`
  Remove the callback function for the memory region.

This `Memcheck` callback extension uses the Ordered Set to record information including the memory starting address, size of the block, callback function pointer, memory operation type, and user callback data. The callback function pointer is a user defined checking routine that will be called when the specified memory operation type is triggered on that registered memory. The callback data is the user defined argument for the callback function. So the callback routine is able to obtain information from the checking tool. The algorithm that manages OSet entries uses binary search with multiple keys. It will search for a match address for the given key in the OSet, and then search the match of the memory size and finally callback function pointer. When the OSet entry is found, the watched memory operation is compared with the current operation, and the callback function could be launched if they match. Special care must be taken to do correct type-casting in the invoked callback, as the user data is always passed as a `void` pointer, it has to be casted correctly, otherwise run-time errors may occur.

6.3.2 MemPin

As Open MPI is supported on Windows platforms, a debugging tool for checking memory errors is also necessary. A new tool named `MemPin` has been designed on top of Intel Pin framework to meet this need.[1]

The `MemPin` tool uses Intel Pin's instrumentation API to provide the same callback functionalities as the `Memcheck` extension for the user application. Furthermore it may be used to perform the basic functionalities of `Memcheck`, such as make memory readable or inacessible, but through a different approach (see Sect. 6.4.2). The available interfaces and descriptions are:

- `MEMPIN_RUNNING_WITH_PIN`
  Checks whether the user application is running under Pin and Pintool.
- `MEMPIN_REG_MEM_WATCH`
  Registers the memory entry for specific memory operation.
- `MEMPIN_UPDATE_MEM_WATCH`
  Updates the memory entry parameters for the specific memory operation.
- `MEMPIN_UNREG_MEM_WATCH`
  Deregisters one memory entry.
- `MEMPIN_SEARCH_MEM_INDEX`
  Returns the memory entry index from the memory address storage.
- `MEMPIN_PRINT_CALLSTACK`
  Prints the current callstack to standard output or a file.

[1]The MemPin tool in this work is developed only targeting at Windows platforms, although it may be used under Linux too.

The user may use the MemPin API to register memory regions with specific callback function and parameter pointers. When the user application is not running with the Pintool, all MemPin calls will be taken as empty macros, and add no overhead. But if running with MemPin tool, Pin first reads the entire executable, and all the MemPin calls will be replaced with the corresponding function calls that are defined inside MemPin. The generated instrumented codes will be then executed and MemPin will observe and respond to the behavior of the user application.

MemPin uses image and trace instrumentations in user applications. The image instrumentation is done when the image is loaded. In this stage, all the MemPin calls used in the user application will be replaced, and the main entry function, like main will be instrumented for starting the trace engine and the callstack log of MemPin.

The next stage will mainly take care of the memory access, callback functions and the user application callstack. The trace instrumentation is analyzed according to each Basic Block (BBL), and every memory operation in the BBL is checked. When the memory is read or written, the single instruction of the memory operation is instrumented with an analysis function with memory information as operands. For generating useful information of where exactly the memory operation has happened, a callstack log engine is instrumented also in this stage. The callstack engine is implemented using a simple C++ stack structure, which stores only the necessary historical instruction addresses of the application and translates the addresses into source information when required. The new function entry address from the caller will be pushed onto the stack, and it is popped off at the end of the callee. To achieve this goal, the tail instruction of each BBL has to be analyzed. More precisely, every "call" and "return" instruction are instrumented for pushing and popping the instruction address stack.

6.4 Memory Checks in Parallel Application

6.4.1 Pre-communication Checks

In Open MPI objects such as communicators, types and requests are declared as pointers to structures. These objects when passed to MPI-calls are being immediately checked for definedness and together with MPI_Status are checked upon exit.[2] Memory being passed to send operations is being checked for accessibility and definedness, while pointers in receive operations are checked for accessibility, only.

[2]E.g. this showed up uninitialized data in derived objects, e.g. communicators created using MPI_Comm_dup.

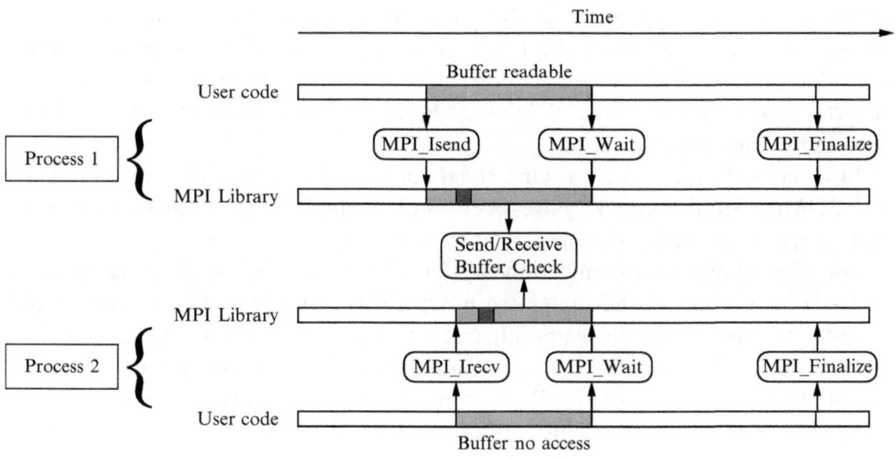

Fig. 6.2 An example of non-blocking communication using pre-communication checks with fragment handling to set accessibility and definedness

Reading or writing to buffers of active, non-blocking receive operations and writing to buffers of active, non-blocking Send-operations are obvious bugs. Buffers being passed to non-blocking operations (after the above checking) are being set to undefined within the MPI-layer of Open MPI until the corresponding completion operation is issued. This setting of the visibility is being set independent of non-blocking `MPI_Isend` or `MPI_Irecv` function. When the application touches the corresponding part in memory before the completion with `MPI_Wait`, `MPI_Test` or multiple completion calls, an error message will be issued. In order to allow the lower-level MPI-functionality to send the user-buffer as fragment, the lower-layer BTLs (Byte Transfer Layers) are adapted to set the fragment in question to accessible and defined, as may be seen in Fig. 6.2. Care has been taken to handle derived datatypes and its implications. Complex datatypes are checked according to their definitions, which means gaps will be ignored to avoid false positive messages.

For send operations, the MPI-1 standard also defines, that the application may not access the send-buffer at all (see [7], p. 30). Many applications do not obey this strict policy, domain-decomposition based applications that communicate ghost-cells, still read from the send-buffer. To the authors' knowledge, no existing implementation requires this policy, therefore the setting to undefined on the Send side is only done when strict-checking is enabled.

For one-sided communications, MPI-2 standard defines that, any conflicting accesses to the same memory location in a window are erroneous (see [8], p. 112). If a location is updated by a put or an accumulate operation, then this location cannot be accessed by a load or another RMA operation until the updating operation is completed on the target. If a location is fetched by a get operation, this location cannot be accessed by other operations as well. When a synchronization call starts, the local communication buffer of an RMA call and a get call should not be updated until it is finished. User buffer of `MPI_Put` or `MPI_Accumulate`, for instance, are set not accessible when these operations are initiated, until the completion operation finished. `Valgrind` will produce an error message, if there is any read or write to the memory area of the user buffer before corresponding completion operation terminates.

In Open MPI, there are two One-sided communication modules, point-to-point and RDMA. Similar checks have been implemented for `MPI_Get`, `MPI_Put`, `MPI_Fence` and `MPI_Accumulate` in point-to-point module.

For the above communication buffer checking, the original features of `Val-grind` are used and integrated in Open MPI on Linux. On the other hand, in order to implement the same checks on Windows, the callback scenario of `MemPin` is used. When the communication starts, for example, the `MPI_Isend` and `MPI_Irecv` are called, all the communication buffers are registered, and all read and write access on the buffers will be checked whether they are legal according to the standards. If an illegal access is found, a warning message with sufficient callstack information will be generated for the user. However, in order to make the `MemPin` efficient and lightweight, memory checks for accessibility and definedness cannot be easily implemented due to complex and large shadow memory consumption.

6.4.2 Post-communication Checking

For more detailed memory checking, one may further implement an interface to encode read and write accesses of previously communicated data. This new analysis mechanism may help user applications to detect whether data has been sent needlessly, i.e. data that has been sent but might be overwritten before reading or may be never accessed in the receiving process.

More precisely, read accesses on the received data is counted as meaningful, while the first write access is not, for the communicated buffer is overwritten before any actual use. To achieve this goal, we make use of the tools introduced in Sects. 6.3.1 and 6.3.2, in order to notify Open MPI of the corresponding tasks based on the type of operations, i.e. read or write to the received buffer.

All the communication buffers are registered when the communication finishes. For example, in Fig. 6.3, when the `MPI_Wait` is called for non-blocking communication, every read and write access on the received buffer will be processed by the callback implementation in Open MPI, and a write before read is marked as illegal. In the finalization phase of the parallel application, all registered memory will be checked for whether there are communicated data but never used, i.e. whether there are buffers without a read access.

Figure 6.4 shows the overall data allocated to store the corresponding information within trivial loop of a receiver. One may see, that the first element in `buf` is first being accessed as read (meaningful), while the second element is accessed first as a write (meaningless) – even following accesses will not change the

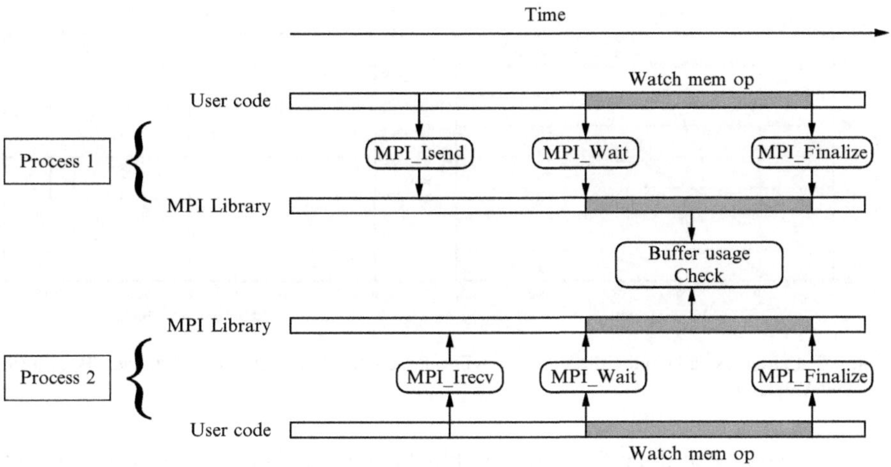

Fig. 6.3 An example of non-blocking communication using post-communication checks

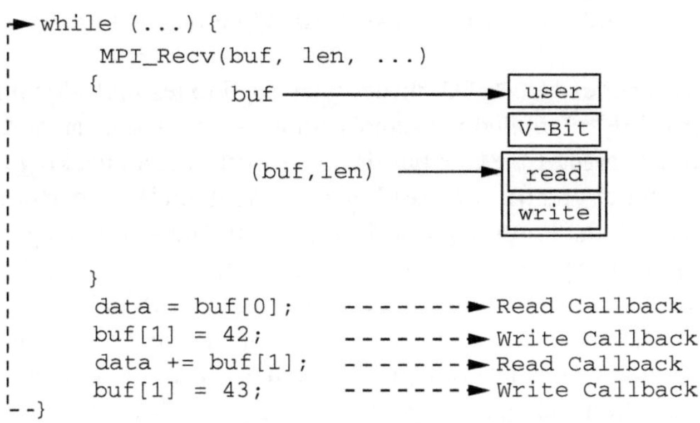

Fig. 6.4 Buffers storing access information of communicated data

outcome of overwriting the communicated data. Upon re-entry into `MPI_Recv`, our extension will trigger a warning of unused data, i.e. data that has been transferred over the wire, but not properly accessed on the receiving side.

6.5 Performance Implications

Adding instrumentation to the code does induce a slight performance hit due to the assembler instructions as explained above, even when the application is not run under `Valgrind`.

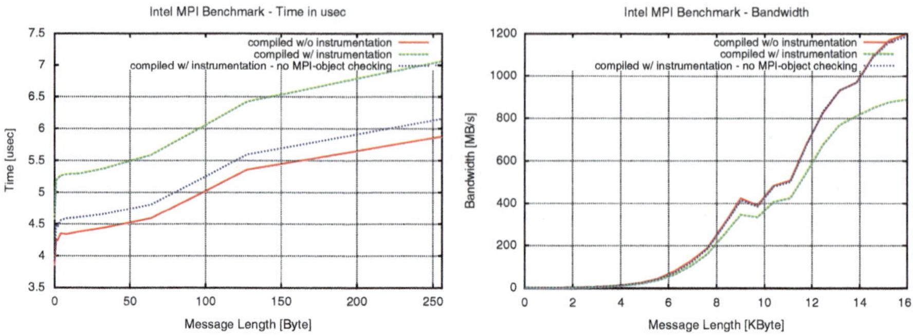

Fig. 6.5 Latencies and bandwidth with and without memchecker instrumentation over IB, running without Valgrind

Tests have been done for both non-blocking communication IMB benchmark, and they were run on the `DGrid`-cluster (Linux) and VISCLUSTER (Windows HPC 2008 R2) at HLRS. Each `DGrid`-cluster machine consists of dual-processor Intel Woodcrest, using Infiniband DDR network with the OpenFabrics stack. And every machine in VISCLUSTER has two AMD Opteron 250 processors and 4.3 GB memory.

For `IMB`, two nodes of `DGrid`-cluster were used to test in following cases: running with Open MPI compiled with and without `--enable-memchecker` and with `--enable-memchecker` but disabled MPI-object checking (see Fig. 6.5) and with and without `Valgrind` (see Fig. 6.6). We include the performance results on two nodes using the PingPong test. In Fig. 6.5 the measured latencies (left) and bandwidth (right) using Infiniband (not running with `Valgrind`) shows the costs incurred by the additional instrumentation, ranging from 18 to 25 % when the MPI-object checking is enabled as well, and 3–6 % when `memchecker` is enabled, but no MPI-object checking is performed. As one may note, while latency is sensitive to the instrumentation added, for larger packet-sizes, it is hardly noticeable anymore (less than 1 % overhead).

Figure 6.6 shows the cost when additionally running with `Valgrind`, again without further instrumentation compared with our additional instrumentation applied, here using TCP connections employing the IPoverIB-interface.

The large slowdown of the MPI-object checking is due to the tests of every argument and its components, i.e. the internal data structures of an `MPI_Comm` consist of checking the definedness of 58 components, checking an `MPI_Request` involves 24 components, while checking `MPI_Datatype` depends on the number of the base types.

On the VISCLUSTER, similar results were seen when running the same benchmark with two nodes over TCP, as shown in Fig. 6.7. The benchmark does not show any performance implications whether the instrumentation is added or not. Of course due to the large memory requirements, the execution shows the expected slow-down when running under `Valgrind`, as every memory access is being checked.

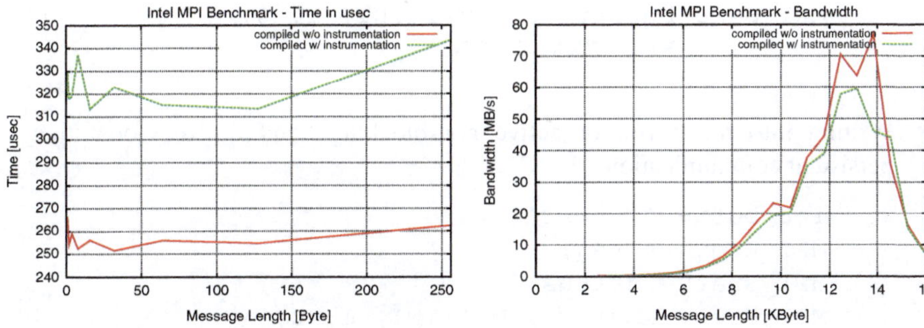

Fig. 6.6 Latencies and bandwidth with and without memchecker instrumentation using IPoverIB, running with Valgrind

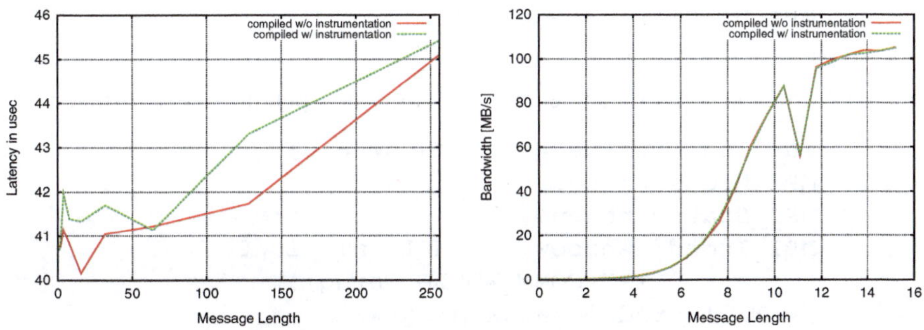

Fig. 6.7 Latencies and bandwidth with and without memchecker instrumentation over TCP on Windows, running without MemPin

6.6 Detectable Error Classes and Findings from Actual Applications

The kind of errors, detectable with a memory debugging tool such as `Valgrind` in conjunction with instrumentation of the MPI implementation are:

- Wrong input parameters, e.g. wrongly sized send buffers:

```
char * send_buffer;
send_buffer = malloc (5);
memset(send_buffer, 0, 5);
MPI_Send(send_buffer, 10, MPI_CHAR, 1, 0, \
         MPI_COMM_WORLD);
```

- Uninitialized input buffers:

```
char * buffer;
buffer = malloc (10);
```

```
MPI_Send(buffer, 10, MPI_INT, 1, 0, \
         MPI_COMM_WORLD);
```

- Writing into the buffer of active non-blocking Send or Recv-operation or persistent communication:

```
int buf = 0;
MPI_Request req;
MPI_Status status;
MPI_Irecv(&buf, 1, MPI_INT, 1, 0, \
          MPI_COMM_WORLD, &req);
/* Will produce a warning */
buf = 4711;
MPI_Wait(&req, &status);
```

- Read from the buffer of active non-blocking Send-operation in strict-mode:

```
int inner_value = 0, shadow = 0;
MPI_Request req;
MPI_Status status;
MPI_Isend(&shadow, 1, MPI_INT, 1, 0, \
          MPI_COMM_WORLD, &req);
/* Will produce a warning */
inner_value += shadow;
MPI_Wait(&req, &status);
```

- Write to the buffer of active get operation:

```
MPI_Win_create(NULL, 0, 1, MPI_INFO_NULL, \
               MPI_COMM_WORLD, &win);
MPI_Win_fence(0, win);
MPI_Get(A, NROWS*NCOLS, MPI_INT, 1, 0, 1, \
        xpose, win);
/* Will produce a warning */
A[1][0] = 4711;
MPI_Win_fence(0, win);
```

- Write before read on received buffer:

```
MPI_Recv(buffer, 2, MPI_CHAR, 1, 0, \
         MPI_COMM_WORLD, &status);
buffer[0] = sizeof(long);
result = buffer[1]*1000;
```

During the course of development, several software packages have been tested with the `memchecker` functionality. Among them problems showed up in Open MPI itself (failed in initialization of fields of the status copied to user-space), an MPI testsuite [3], where tests for the `MPI_ERROR` triggered an error. In order to reduce the number of false positives in Infiniband networks, the `ibverbs` library of the OFED stack [13] was extended with instrumentation for buffer passed back from kernel-space.

6.7 Conclusion

We have presented an implementation of memory debugging features into Open MPI, using the instrumentation of the `Valgrind` and a newly developed Intel Pintool, and the performance implication of using the instrumentation with several benchmarks. This allows detection of hard-to-find bugs in MPI parallel applications, libraries and Open MPI itself [2]. Up to now, no other debugger is able to find these kinds of errors.

With regard to related work, debuggers such as Umpire [15], Marmot [5] or the Intel Trace Analyzer and Collector [2], actually any other debugger based on the Profiling Interface of MPI, may detect bugs regarding non-standard access to buffers used in active, non-blocking communication without hiding false positives of the MPI-library itself.

Acknowledgements This work was funded by project Int.EU.Grid (Interactive EUropean Grid) with EU-Contract Number 031857, by project DORII (Deployment of Remote Instrumentation Infrastructure) with EU-Contract Number 213110 and by Microsoft Technical Computing Initiative (TCI) since March 2007.

We would like to thank Julian Seward and the open source community for `valgrind`, which has proven invaluable in many software projects.

References

1. Bcheck Man Page from SUN Developers Website. Internet. http://developers.sun.com/sunstudio/documentation/ss11/mr/man1/bcheck.1.html (2011)
2. DeSouza, J., Kuhn, B., de Supinski, B.R.: Automated, scalable debugging of MPI programs with Intel message checker. In: Proceedings of the 2nd International Workshop on Software Engineering for High Performance Computing System Applications, vol. 4, pp. 78–82. ACM Press, New York (2005)
3. Keller, R., Resch, M.: Testing the correctness of MPI implementations. In: Proceedings of the 5th International Symposium on Parallel and Distributed Computing Conference, Timisoara, pp. 291–295 (2006)
4. Keller, R., Fan, S., Resch, M.: Memory debugging of MPI-parallel applications in open MPI. In: Joubert, G., Bischof, C., Peters, F., Lippert, T., Bucker, M., Gibbon, P., Mohr B. (eds.) Proceedings of ParCo'07, Julich (2007)

5. Krammer, B., Mueller, M.S., Resch, M.M.: Runtime checking of MPI applications with Marmot. In: Proceedings of the International Conference ParCo 2005, Malaga (2005)
6. Luk, C.K., Cohn, R., Muth, R., Patil, H., Klauser, A., Lowney, G., Wallace, S., Reddi, V.J., Hazelwood, K.: Pin: building customized program analysis tools with dynamic instrumentation. In: Proceedings of the 2005 ACM SIGPLAN Conference on Programming Language Design and Implementation, pp. 190–200. ACM, New York (2005)
7. Message Passing Interface Forum: MPI: A Message Passing Interface Standard. http://www.mpi-forum.org (1995)
8. Message Passing Interface Forum: MPI-2: Extensions to the Message-Passing Interface. http://www.mpi-forum.org (1997)
9. MPI Forum Ticket Number 45. Internet. https://svn.mpi-forum.org/trac/mpi-forum-web/ticket/45 (2008)
10. Seward, J., Nethercote, N.: Using Valgrind to detect undefined value errors with bit-precision. In: Proceedings of the USENIX'05 Annual Technical Conference, Anaheim (2005)
11. Shiqing Fan, R.K., Resch, M.: Enhanced memory debugging of mpi-parallel applications in open mpi. In: 4th Parallel Tools Workshop, Stuttgart (2010)
12. Srivastava, A., Eustace, A.: Atom: A System for Building Customized Program Analysis Tools, pp. 196–205. ACM, New York (1994)
13. The Open Fabrics Project Webpage. https://www.openfabrics.org (2007)
14. Totalview Memory Debugging capabilities. http://www.etnus.com/TotalView/Memory.html
15. Vetter, J.S., de Supinski, B.R.: Dynamic software testing of MPI applications with umpire. In: Proceedings of Supercomputing (SC), Dallas. http://www.sc2000.org/proceedings/techpapr/index.htm (2000)
16. Woodall, T., Graham, R., Castain, R., Daniel, D., Sukalski, M., Fagg, G., Gabriel, E., Bosilca, G., Angskun, T., Dongarra, J., Squyres, J., Sahay, V., Kambadur, P., Barrett, B., Lumsdaine, A.: Open MPI's TEG point-to-point communications methodology: comparison to existing implementations. In: Recent Advances in Parallel Virtual Machine and Message Passing Interface, vol. 3241, pp. 105–111. Springer, Budapest (2004)

Chapter 7
Score-P: A Joint Performance Measurement Run-Time Infrastructure for Periscope, Scalasca, TAU, and Vampir

Andreas Knüpfer, Christian Rössel, Dieter an Mey, Scott Biersdorff, Kai Diethelm, Dominic Eschweiler, Markus Geimer, Michael Gerndt, Daniel Lorenz, Allen Malony, Wolfgang E. Nagel, Yury Oleynik, Peter Philippen, Pavel Saviankou, Dirk Schmidl, Sameer Shende, Ronny Tschüter, Michael Wagner, Bert Wesarg, and Felix Wolf

Abstract This paper gives an overview about the Score-P performance measurement infrastructure which is being jointly developed by leading HPC performance tools groups. It motivates the advantages of the joint undertaking from both the developer and the user perspectives, and presents the design and components of the newly developed Score-P performance measurement infrastructure. Furthermore, it contains first evaluation results in comparison with existing performance tools and presents an outlook to the long-term cooperative development of the new system.

7.1 Introduction

The HPC community knows many established or experimental tools that assist programmers with the performance analysis and optimization of parallel target applications. Well known examples are HPCToolkit [1], Jumpshot [20], Paraver [12], Periscope [2], Scalasca [8], TAU [16], and Vampir [11]. They form an important contribution to the HPC ecosystem, because HPC is naturally obsessed with good performance and high efficiency, on one hand, and with more and more complex hardware and software systems, on the other hand – in this situation, sophisticated tools are an absolute necessity for analyzing and optimizing complicated parallel performance behavior.

A. Knüpfer (✉) · C. Rössel · D. an Mey · S. Biersdorff · K. Diethelm · D. Eschweiler · M. Geimer · M. Gerndt · D. Lorenz · A. Malony · W.E. Nagel · Y. Oleynik · P. Philippen · P. Saviankou · D. Schmidl · S. Shende · R. Tschüter · M. Wagner · B. Wesarg · F. Wolf
ZIH, TU Dresden, 01062 Dresden, Germany
e-mail: andreas.knuepfer@tu-dresden.de

H. Brunst et al. (eds.), *Tools for High Performance Computing 2011*,
DOI 10.1007/978-3-642-31476-6_7, © Springer-Verlag Berlin Heidelberg 2012

There are several established performance tools that co-exist today and many more experimental ones. They focus on different aspects and provide complementary specialized features, which makes it worthwhile to use them in combination. But at the same time, there are many similarities and overlapping functionality, in particular redundant basic functionality. This is almost necessarily so, because there is only a small number of options for several aspects, for example:

- Parallelization models (e.g. message passing, multi-threading, PGAS),
- Performance data acquisition (e.g. instrumentation, sampling, assisted by hardware components),
- Performance data representation (events, statistics).

Usually, a particular tool covers multiple combinations from this list. Therefore, one finds many redundancies in existing tools that serve the same purpose. Typical examples are source code instrumentation, profiling or event recording for MPI, and data formats. Often, one implementation could replace the other in terms of functionality and only the development histories of different tools prevent them from (re-)using the same components.

We argue that this fragmentation in the performance tools landscape brings some disadvantages for both groups concerned, the developers and the users. And we are convinced that the joint approach presented in this paper will be of mutual benefit.

7.1.1 Motivation for a Joint Measurement Infrastructure

The redundancies found in the tools landscape today bring a number of disadvantages for the developers as well as the users of the tools. From the tool developers perspective, redundancies are mostly found in the data acquisition parts, such as instrumentation, data acquisition from various sources, and in the collected data and data formats. The actual data evaluation approaches are diverse enough to justify separate tools. Thus, our goal is a joint measurement infrastructure that combines the similar components and interfaces to diverse tools. This will save redundant effort for development, maintenance, porting, scalability enhancement, support, and training in all participating groups. The freed resources will be available for enhancing analysis functionality in the individual tools.

From the user perspective, the redundancies in separated tools are a discomfort because it requires multiple learning curves for different measurement systems. Incompatible configurations, different options for equivalent features, missing features, etc. makes it harder to switch between tools or combine them in a reproducible manner. Since parallel programming and performance analysis at large scales are by no means trivial, the fragmentation in tools adds to the existing challenge for the users. The same is true for incompatible data formats that in principle transport the same data. At best, unnecessary data conversion is required, at the worst this causes repeated measurements of the same situations with different tools. Both can become very expensive for today's highest-scale use cases. Last but not least, installation and updates of the software packages require redundant effort on the user side. A joint infrastructure will remove many of those obstacles and improve interoperability.

In the following section, the paper introduces the SILC and PRIMA projects as the frame of all presented work. The main components of the joint infrastructure are discussed in Sects. 7.3–7.7, followed by early evaluations of the run-time measurement component and the event trace data format in Sect. 7.8. The paper ends with an outlook to future goals beyond the projects' funding periods and the conclusions.

7.2 The SILC and PRIMA Projects

The SILC project[1] ("Skalierbare Infrastruktur zur Automatischen Leistungsanalyse Paralleler Codes", Engl. "Scalable Infrastructure for Automatic Performance Analysis of Parallel Codes") and the PRIMA project[2] ("Performance Refactoring of Instrumentation, Measurement, and Analysis Technologies for Petascale Computing") bring together the following established performance tools groups:

- The Scalasca groups from Forschungszentrum Jlich, Germany and the German Research School for Simulation Sciences, Aachen, Germany,
- The Vampir group at Technische Universitt Dresden, Germany,
- The Persicope group at Technische Universitt Mnchen, Germany, and
- The TAU group at University of Oregon, Eugene, USA,

and as further partners

- The Computing Center at RWTH Aachen, Germany, and
- The GNS mbH, Braunschweig, Germany, as industry partner.

All partners have a long history of mutual cooperation. The projects' main goal is to provide a joint infrastructure to address the disadvantages of separate tools as discussed above. The new infrastructure created in SILC and PRIMA has to support all central functionalities that were present in the partner's tools beforehand:

Periscope is an on-line analysis tool based on profiling information. During run-time it evaluates performance properties and tests hypotheses about typical performance problems, which are reported to the user if detected.

Scalasca performs post-mortem analysis of event traces and automatically detects performance critical situations. The results are categorized and hierarchically arranged according to types of performance problems, source code locations, and physical (computing hardware) or logical (computation domain) location.

[1]The SILC project is funded by the BMBF Germany under grant number 01IH08006.

[2]The PRIMA project is funded by the US DOE under grant number DE-SC0001621.

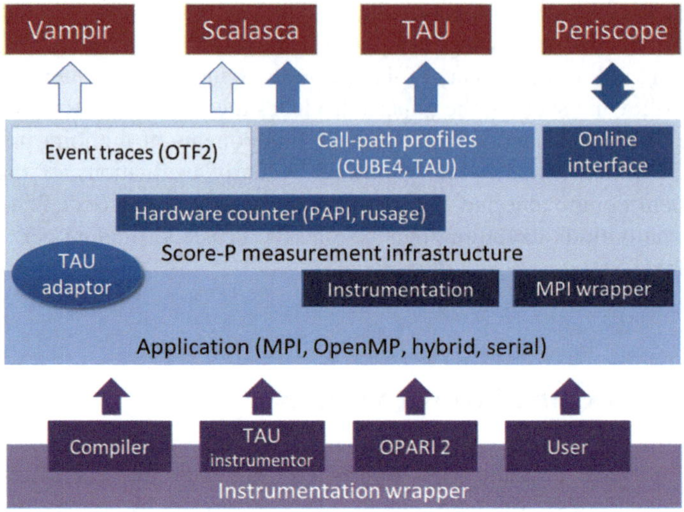

Fig. 7.1 Architecture of the Score-P instrumentation and measurement system with connection points to the currently supported analysis tools

TAU is a extensive profiling tool-set, which allows to collect various kinds of profiles such as flat profiles, phase profiles, and call-path profiles. Furthermore, it provides flexible visualization and correlation of performance data.

Vampir is an interactive event trace visualization software which allows to analyze parallel program runs with various graphical representations in a post-mortem fashion.

A number of design goals were formulated to address the needs of the four tools as well as new tools by other groups in the future. On the one hand, the primary functional requirements are:

- Support profiling and event tracing,
- Initially use direct instrumentation, later also add sampling,
- Allow off-line access (post-mortem analysis) and on-line access (live analysis),
- Initially target parallelization via MPI 2.1 and OpenMP 3.0 as well as hybrid combinations, later also include CUDA, OpenCL, and others.

On the other hand, the following non-functional requirements were considered:

- Portability to all major HPC platforms and robustness,
- Scalability to petascale level,
- Low measurement overhead,
- Easy single-step installation (through the UNITE framework),
- Open source with New BSD license.

From those goals and the experience from the existing tools, the architecture shown in Fig. 7.1 was derived. It contains the high-level components Score-P, OTF2, CUBE4, and OPARI2 which are presented in detail in the following sections.

7.3 Score-P

The Score-P component consists of an instrumentation framework, several run-time libraries, and some helper tools. The instrumentation allows users to insert measurement probes into C/C++ and Fortran codes that collect performance-related data when triggered during measurement runs. In order to collect relevant data, e.g., times, visits, communication metrics, or hardware counters, the applications need to link against one of several provided run-time libraries for serial execution, OpenMP or MPI parallelism or hybrid combinations. Extensions for POSIX threads, PGAS and CUDA are planned. The collected performance data can be stored in the OTF2, CUBE4, or TAU snapshot formats or queried via the on-line access interface.

Let's look at the instrumentor command *scorep* in more detail. It is used as a prefix to the usual compile and link commands, i.e., instead of `mpicc -c foo.c` one uses the command `scorep mpicc -c foo.c`. The instrumentor detects the programming paradigm used, MPI in this case, and adds appropriate compiler and linker flags before executing the actual build command. Currently, the following means of instrumentation are supported:

- Compiler instrumentation,
- MPI library interposition,
- OpenMP source code instrumentation using OPARI2 (see Sect. 7.6),
- Source code instrumentation via the TAU instrumentor [6], and
- User instrumentation using convenient macros.

These instrumentation variants are configurable by options to the `scorep` command. It is also possible to use the Score-P headers and libraries directly. In this case, the helper tool `scorep-config` provides all necessary information.

Once the application is instrumented, the user just needs to run it in order to collect measurement data. In contrast to the predecessor measurement systems, a flexible and efficient memory management system was implemented that stores data in thread-local chunks of pre-allocated memory. This allows an efficient distribution of the available memory to an arbitrary number of threads and minimizes the perturbation and overhead due to run-time memory allocations.

By default, Score-P runs in profiling mode and produces data in the CUBE4 format (see Sect. 7.5). This data provides first insight about the hot-spots in the application and allows to customize options for subsequent measurement runs, for example, to select MPI groups to record, to specify the total amount of memory the measurement is allowed to consume, to adapt trace output options, to specify a set of hardware counters to be recorded, and more. To keep the amount of measurement data manageable, one can include or exclude regions by name and/or wild-cards (filtering). In tracing mode, one might restrict the recording to specific executions of a region (selective tracing). In the end, the recorded data (either profiling or tracing data or both) is written to a uniquely named experiment directory. When switching from profiling to tracing or on-line access for successive experiments, there is no need to recompile or re-instrument the target application.

Furthermore, two important scalability limitations of former generations of tools have been addressed. The first affected codes using many MPI communicators, which were not handled adequately at very large scales. A new model according to [9] solved this for Score-P. The second affected the unification of identifiers at the very end of measurement. It generates globally unique identifiers from the local identifiers used during data recording. In the past, an algorithm with linear complexity with respect to the number of parallel processes was used. Obviously, this cannot match today's and future scalability demands. The new implementation uses a tree-based reduction as presented in [7]. With these fundamental improvements, Score-P is fit for the scalability levels of future supercomputer generations.

7.4 The Open Trace Format Version 2

The Open Trace Format Version 2 (OTF2) is a newly designed software package based on the experiences of the two predecessor formats OTF1 [10] and EPILOG [19], the native formats of Vampir and Scalasca, respectively.

Its main characteristics are similar to other record-based parallel event trace formats. It contains events and definitions and distributes data storage over multiple files. It uses one anchor file, one global definition file, n local definition files, and n local event files for n processes or threads. Inside each event file, the event records are stored in temporal order. The essence of OTF2 is not only the format specification but also a read/write API and a library with support for selective access. OTF2 is available as a separate software package, but also a central part of the Score-P run-time system where it acts as the memory buffer for event trace collection. It uses the same binary encoding for the memory buffers and the file representation.

As a special feature, OTF2 supports multiple I/O substrates – these are exchangeable back-ends that allow different treatment of the I/O. The standard substrate stores the OTF2 memory buffer representation in multiple files. As an alternative, a compression substrate uses ZLib for data compression of the individual files. A future extension plans to pass persistent memory pages from the trace processes or threads to post-mortem analysis processes and thus avoid I/O altogether. Last but not least, the SION library [4] substrate will address the challenge of massively parallel data and meta-data requests which overcharge parallel file systems. This is a fundamental problem on highest-scale HPC machines today. For more details about OTF2 in general see [3].

7.4.1 The SIONlib OTF2 Substrate

OTF2 distributes the records for each thread or process into separate files to avoid additional synchronization during measurement. While this is not problematic for lower scales (e.g., smaller than 1,000 ranks), it becomes a severe problem for parallel file systems at large scales. The main reason is insufficient scalability of meta-data handling in parallel file systems. The write bandwidth is usually no limitation.

The problem with huge amounts of file handles is a widely known issue also for other applications in high-performance computing. SIONlib is a parallel I/O library which was developed at Forschungszentrum Jülich to overcome this problem by mapping virtual file handles onto a single physical file. This approach has been shown to scale up to several thousands of processes [4] and is, therefore, suitable to enhance the scalability of OTF2. In OTF2, the SION library is integrated as an additional I/O substrate which produces two separate multi-files, one for all definitions and one for all events.

7.5 The CUBE4 Format and GUI

CUBE is a profiling data model and file format. The data model describes the performance behavior of an application along three dimensions. The first dimension is a set of performance metrics. They characterize the kind of the performance issues under consideration. The second dimension is the call tree, which shows the place in the application execution where a certain issue appears. The third dimension is a description of the system which executed the application. Each triple of coordinates in this three-dimensional performance space is mapped onto a numeric value, which represents the severity of a given performance metric while visiting a given call path at a given system location.

To store measured profile data, Score-P uses the CUBE4 framework [7]. Like its predecessor CUBE3, CUBE4 provides a set of libraries and tools to store and analyze performance profiles. It consists of a writer library designed for scalability, a general-purpose C++ reader and re-writer library, a set of tools for manipulating measured profiles, and a graphical user interface for visual inspection of profiles. In contrast to its predecessor, CUBE4 also provides a Java reader library. The data format differs from CUBE3 in that the severity values, which constitute the bulk of the data, are now stored in a binary format. XML is retained only for the metadata part, since using XML for everything turned out to be too inefficient when large-scale data had to be written. GUI response times are improved using techniques such as dynamic loading, inclusive storage, and sparse representation of the data.

7.6 The OPARI2 Instrumentor for OpenMP

The source-to-source instrumentor OPARI is used to automatically wrap OpenMP constructs like parallel regions with calls to the performance monitoring interface POMP [14]. OPARI is used in many performance tools like Scalasca, Vampir and ompP [5]. In order to support version 3.0 of the OpenMP specification [15], OPARI was enhanced to support OpenMP tasking and to provide POMP implementers with information for OpenMP nesting.

The new way to exploit parallelism in OpenMP programs using tasks posed two critical challenges for event-based analysis tools. Firstly, with OpenMP tasking, where tasks can be suspended and resumed, it is no longer guaranteed that the order of region enter and exit events follows a strict LIFO semantics per thread (the call stack). The solution to this problem is to distinguish individual task instances and to track their suspension and resumption points [13].

Secondly, with OpenMP nesting the number of threads in a parallel region is a priori unknown and the thread IDs are no longer unique, whereas performance tools traditionally rely on pre-allocated buffers for a fixed set of threads indexed by the thread IDs. Therefore, OPARI2 provides an upper bound for the number of threads used in the next parallel region as assistance for the run-time system.

In addition, OPARI2 comes with an enhanced linking procedure that allows to keep all instrumentation information within the compilation units themselves. Additional index files are no longer needed, which improves the usability of OPARI2 for multi-directory builds and pre-compiled libraries. With the presented improvements, the new major version OPARI2 addresses all aspects of state-of-the-art parallelization with OpenMP, either alone or in combination with MPI.

7.7 The On-Line Access Interface

The On-line Access interface offers a remote access path to the rich profiling capabilities of Score-P. It allows to configure measurement parameters, retrieve profile data, and to interrupt and resume the application execution. This functionality enables tools for iterative on-line performance analysis.

The On-line Access interface accepts socket connections from a remote analysis tool and communicates via the extensible and easy to use text-based Monitoring Request Interface (MRI) protocol which supports three classes of requests:

- *Monitoring requests* to select performance metrics to be recorded and to adjust measurement parameters,
- *Data retrieval requests* to fetch selected performance data, and
- *Application execution control requests* to control the measurement window based on predefined suspension points during application execution.

The main advantage of the Score-P On-line Access is that on-line analysis tools can decouple measurement from analysis, which reduces perturbation of the results. Furthermore, profile data don't need to be stored to files but can be consumed by analysis components directly. One example is the Periscope on-line analysis tool [2] which queries profile data for pre-selected phases of an iterative parallel application. Based on this, it derives hypotheses about performance problems and refines the measurement for following phases to prove or disprove them.

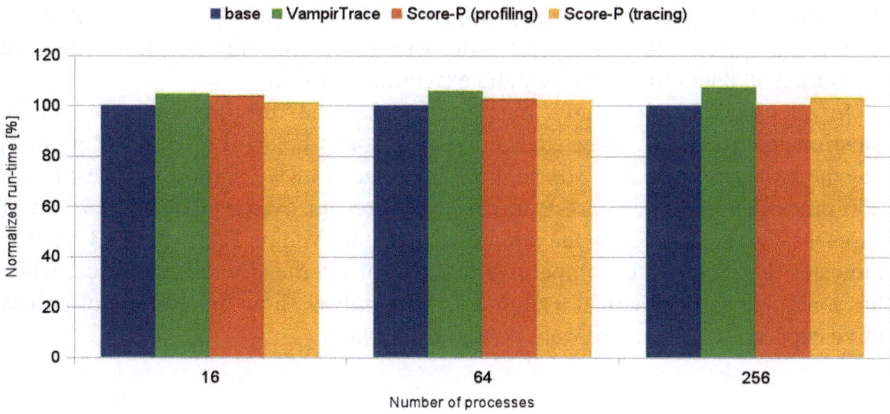

Fig. 7.2 Comparison of the run-times of the COSMO code in different modes: without instrumentation (base), with VampirTrace event tracing, with Score-P profiling, and with Score-P tracing. All times are relative to the uninstrumented run

7.8 Early Evaluation

Minimal overhead and resource consumption are essential requirements for the Score-P infrastructure. Therefore, the run-time overhead of the Score-P measurement system was compared to the existing VampirTrace infrastructure and the memory consumption of OTF2 was compared to both well-established predecessor trace formats OTF and EPILOG and the internal representation in VampirTrace.

7.8.1 Run-Time Measurement Overhead

As initial evaluation of the run-time overhead, the Score-P measurement system was tested with the COSMO code using 16, 64, and 256 MPI processes on an SGI Altix 4700. COSMO is an atmospheric model code mainly developed by the DWD[3] [17].

The baseline of this comparison are executions of the uninstrumented COSMO code. In addition, the application was executed with VampirTrace event tracing and with Score-P both in profiling and tracing mode. Figure 7.2 shows the resulting run-times. All values are normalized to the uninstrumented case and contain only

[3]Deutscher Wetterdienst (German Meteorological Service), see also http://www.cosmo-model.org/content/model/general/default.htm

the measurement phase excluding post-processing or output of measurement results to the file system. In the typical tools workflow, this indicates how close the measurement data reflects the original execution situation.

In all cases, Score-P only slightly increases the run-time compared to the unistrumented case as expected; the overhead stays below 4 %. Score-P profiling has the largest observed overhead at 16 processes (3.8 %) and the smallest one at 256 processes (0.2 %). Score-P tracing generates an overhead of 1–3.2 %. In all cases the overhead is below the overhead caused by VampirTrace (4.6–7.3 %). This indicates that Score-P is competitive in terms of overhead to VampirTrace, which has long proven its practical usefulness. All numbers show the average of at least three experiments and there were no notable outliers.

7.8.2 Trace Format Memory Consumption

Secondly, the memory consumption for event trace data is evaluated. The OTF2 data representation, which is identical to the Score-P memory buffer representation, is compared to the EPILOG format, the OTF (version 1) format, and VampirTrace's memory representation. This evaluation reveals how long the event recording can be continued with a given buffer size before it has to be interrupted for writing out data. As examples the following applications and benchmarks are used:

- The SPEC MPI2007[4] benchmarks with the test cases 104.milc ... 137.lu
- The previously mentioned COSMO code[5] from DWD
- The NAS Parallel Benchmarks[6] with nas_pb_bt (block tridiagonal solver)
- The SMG2000 Benchmark[7] (semicoarsening multigrid solver)
- The ASCI SWEEP3D benchmark[8](3D discrete neutron transport)

Figure 7.3 shows a comparison of the memory consumption of VampirTrace as the baseline, OTF, EPILOG, and OTF2. The results show that OTF2 and Score-P perform better than both predecessor tool sets and the situation behaves very similar for all test cases. OTF2 reduces the memory consumption by about 70 % compared to VampirTrace, 20–35 % compared to OTF, and 14–17 % compared to EPILOG.

[4]http://www.spec.org/mpi/

[5]http://www.cosmo-model.org/content/model/general/default.htm

[6]http://www.nas.nasa.gov/Resources/Software/npb.html

[7]https://asc.llnl.gov/computing_resources/purple/archive/benchmarks/smg/

[8]http://www.ccs3.lanl.gov/pal/software/sweep3d/

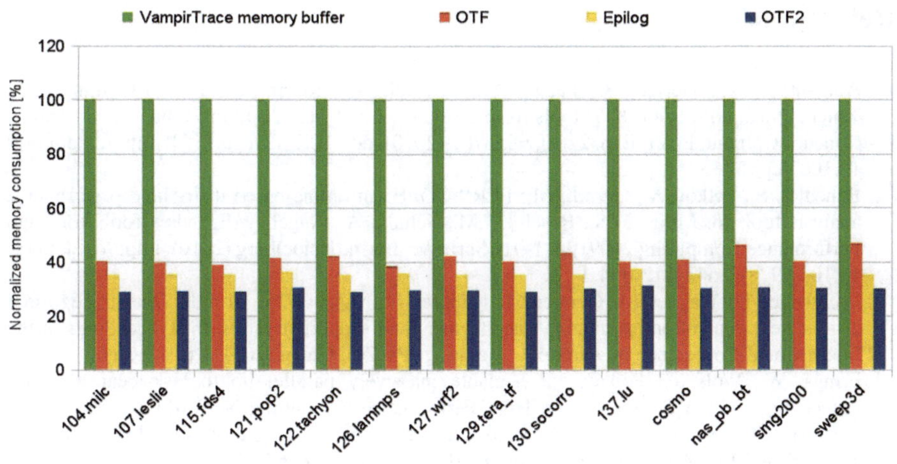

Fig. 7.3 Comparison of the event trace data memory consumption of VampirTrace, OTF, EPILOG, and OTF2 for a series of experiments with typical HPC benchmark codes

7.9 Conclusion and Outlook

With Score-P, the user community will have a single platform for large-scale performance measurements that can be interpreted with a rich collection of analysis tools. This is a first big step towards integrating the so far fragmented performance tools landscape. The primary benefits for the user are simplified installation of our tool suite, as only one measurement system has to be installed, and reduced learning effort, as the user has to read only one set of measurement instructions. Less obvious but equally important, eliminating redundancy among individual tools will free substantial tool development resources that can from now on be redirected to more powerful analysis features. At the time of writing, version 1.0 of the Score-P tool set is going to be released.

While the objective of the projects SILC and PRIMA has been the creation of an initial version, the software is already subject of a number of ongoing follow-up projects. The European project H4H (2010–2013) in the ITEA-2 framework will add extensions needed for heterogeneous architectures to Score-P with funding from the German Ministry of Education and Research (BMBF). Capabilities for compression of time-series call-path profiles [18], a feature to study the dynamic performance behavior of long-running codes, will be integrated in the EU/FP7 project HOPSA. Finally, the BMBF project LMAC will not only further refine this method but also provide functionality for the semantic compression of event traces, a prerequisite for analyzing traces of long-running applications more effectively. In addition to adding new technical features, it will also establish rules that govern Score-P's further evolution, striking a balance between robustness and stability on the one hand and innovation on the other.

References

1. Adhianto, L., Banerjee, S., Fagan, M., Krentel, M., Marin, G., Mellor-Crummey, J., Tallent, N.R.: HPCTOOLKIT: tools for performance analysis of optimized parallel programs. Concurr. Comput. Pract. Exp. **22**(6), 685–701 (2010). doi: 10.1002/cpe.1553. http://dx.doi.org/10.1002/cpe.1553
2. Benedict, S., Petkov, V., Gerndt, M.: PERISCOPE: an online-based distributed performance analysis tool. In: Mller, M.S., Resch, M.M., Schulz, A., Nagel, W.E. (eds.) Tools for High Performance Computing 2009, pp. 1–16. Springer, Berlin/Heidelberg (2010). http://dx.doi.org/10.1007/978-3-642-11261-4_1
3. Eschweiler, D., Wagner, M., Geimer, M., Knüpfer, A., Nagel, W.E., Wolf, F.: Open trace format 2 – the next generation of scalable trace formats and support libraries. In: Proceedings of the International Conference on Parallel Computing (ParCo), Ghent (2011). (to appear)
4. Frings, W., Wolf, F., Petkov, V.: Scalable massively parallel i/o to task-local files. In: Proceedings of the Conference on High Performance Computing Networking, Storage and Analysis, SC '09, pp. 17:1–17:11. ACM, New York, NY (2009). doi: http://doi.acm.org/10.1145/1654059.1654077. http://doi.acm.org/10.1145/1654059.1654077
5. Fürlinger, K., Gerndt, M.: ompP – a profiling tool for OpenMP. In: 1st International Workshop, IWOMP 2005, Eugene, OR, USA, June 1–4, 2005. LNCS 4315, Springer
6. Geimer, M., Shende, S.S., Malony, A.D., Wolf, F.: A generic and configurable source-code instrumentation component. In: ICCS 2009: Proceedings of the 9th International Conference on Computational Science, pp. 696–705. Springer, Berlin (2009)
7. Geimer, M., Saviankou, P., Strube, A., Szebenyi, Z., Wolf, F., Wylie, B.J.N.: Further improving the scalability of the Scalasca toolset. In: Proceedings of PARA 2010: State of the Art in Scientific and Parallel Computing, Minisymposium Scalable Tools for High Performance Computing, Reykjavik. Springer, Berlin (2010)
8. Geimer, M., Wolf, F., Wylie, B.J., Ábrahám, E., Becker, D., Mohr, B.: The scalasca performance toolset architecture. Concurr. Comput. Pract. Exp. **22**(6), 702–719 (2010). doi: 10.1002/cpe.1556
9. Geimer, M., Hermanns, M.A., Siebert, C., Wolf, F., Wylie, B.J.N.: Scaling performance tool MPI communicator management. In: Proceedings of the 18th European MPI Users' Group Meeting (EuroMPI), Santorini. Lecture Notes in Computer Science, vol. 6960, pp. 178–187. Springer, Berlin (2011). doi: 10.1007/978-3-642-24449-0_21
10. Knüpfer, A., Brendel, R., Brunst, H., Mix, H., Nagel, W.E.: Introducing the open trace format (OTF). In: Computational Science ICCS 2006: 6th International Conference, LNCS 3992. Springer, Reading (2006)
11. Knüpfer, A., Brunst, H., Doleschal, J., Jurenz, M., Lieber, M., Mickler, H., Müller, M.S., Nagel, W.E.: The vampir performance analysis tool set. In: Resch, M., Keller, R., Himmler, V., Krammer, B., Schulz, A. (eds.) Tools for High Performance Computing, pp. 139–155. Springer, Berlin (2008)
12. Labarta, J., Gimenez, J., Martínez, E., González, P., Harald, S., Llort, G., Aguilar, X.: Scalability of tracing and visualization tools. In: Joubert, G.R., Nagel, W.E., Peters, F.J., Plata, O.G., Tirado, P., Zapata, E.L. (eds.) Parallel Computing: Current and Future Issues of High-End Computing, Proceedings of the International Conference ParCo 2005, Jülich, 13–16 Sept 2005. Department of Computer Architecture, University of Malaga, Spain, John von Neumann Institute for Computing Series, vol. 33, pp. 869–876. Central Institute for Applied Mathematics, Jülich (2005)
13. Lorenz, D., Mohr, B., Rössel, C., Schmidl, D., Wolf, F.: How to reconcile event-based performance analysis with tasking in OpenMP. In: proceedings of 6th International Workshop on OpenMP (IWOMP), Tsukuba. LNCS, vol. 6132, pp. 109–121. Springer, Berlin (2010)
14. Mohr, B., Malony, A.D., Shende, S., Wolf, F.: Design and prototype of a performance tool interface for OpenMP. J. Supercomput. **23**(1), 105–128 (2002)

15. OpenMP Architecture Review Board: OpenMP application program interface, Version 3.0. http://www.openmp.org/mp-documents/spec30.pdf
16. Shende, S., Malony, A.D.: The TAU parallel performance system, SAGE publications. Int. J. High Perform. Comput. Appl. **20**(2), 287–331 (2006)
17. Steppeler, J., Doms, G., Schttler, U., Bitzer, H.W., Gassmann, A., Damrath, U., Gregoric, G.: Meso-gamma scale forecasts using the nonhydrostatic model lm. Meteorol. Atmos. Phys. **82**, 75–96 (2003). http://dx.doi.org/10.1007/s00703-001-0592-9. 10.1007/s00703-001-0592-9
18. Szebenyi, Z., Wolf, F., Wylie, B.J.N.: Space-efficient time-series call-path profiling of parallel applications. In: Proceedings of the ACM/IEEE Conference on Supercomputing (SC09), Portland. ACM, New York (2009)
19. Wolf, F., Mohr, B.: EPILOG binary trace-data format. Technical Report FZJ-ZAM-IB-2004-06, Forschungszentrum Jülich (2004)
20. Wu, C.E., Bolmarcich, A., Snir, M., Wootton, D., Parpia, F., Chan, A., Lusk, E., Gropp, W.: From trace generation to visualization: a performance framework for distributed parallel systems. In: Proceedings of SC2000: High Performance Networking and Computing, Dallas (2000)

Chapter 8
Trace-Based Performance Analysis for Hardware Accelerators

Guido Juckeland

Abstract Hardware accelerators have changed the HPC landscape as they open a potential route to exascale computing. At the same time they also complicate the tasl of application development since they introduce another level of parallelism and, thus, complexity. Performance tool support to aid the developer will be a necessity. While profiling is offered by the accelerator vendors, tracing tools can also adopt hardware accelerators. A number of challenges for data acquisition and visualization and their solutions are presented in this paper.

8.1 Introduction

As the multi-core era is embracing many-core architectures, high performance computing (HPC) is also gaining an increased momentum using heterogeneous processing elements. This results in another dimension of parallel computing being added—apart from message passing and multi-threading—to the already complex programming ecosystem to harness the capabilities of the heterogeneous hardware. Balancing the compute load over the three levels of parallelism is a delicate task which is assisted by tools that can report information about the application's performance. While vendor tools for hardware accelerators very well cover the parallelism at the hardware accelerator level, they usually fail to cover the other two parallelism levels. The opposite holds for HPC performance tools.

After a brief introduction of performance tools for hardware accelerators this paper presents the extension of the well renowned tool VampirTrace to capture also accelerator activity. A general approach for trace-based performance analysis for

G. Juckeland (✉)
Center for Information Services and High Performance Computing (ZIH), Technische Universität Dresden, 01062 Dresden, Germany
e-mail: guido.juckeland@tu-dresden.de

H. Brunst et al. (eds.), *Tools for High Performance Computing 2011*,
DOI 10.1007/978-3-642-31476-6_8,

hardware accelerated application is introduced, as well as specific modifications to cover the current most widely used accelerator paradigms, CUDA and OpenCL. Furthermore, it discusses how the additional hardware accelerator related data affect the trace format and data visualization.

8.2 Related Work

The hardware accelerator ecosystems from any vendor includes besides the accelerator hardware, also a software environment usually consisting of a compiler, a debugger and some type of performance analysis tool. The NVIDIA CUDA environment includes the Visual Profiler [13] which is available on all CUDA supported platforms and NVIDIA Parallel NSight [14] which can be integrated into Microsoft's Visual Studio. OpenCL accelerated applications running on AMD APUs as well as AMD Radeon GPUs can be profiled with AMD's APP Profiler [1]. The tools from both vendors allow profiling of CUDA or OpenCL accelerated applications even if they are multi-threaded. They do not, however, allow analysis of message passing based accelerated applications. Other less common accelerator platforms offer similar tools: the IBM Cell/B.E. offers profiling and tracing tools for its architecture [8], and the Clearspeed accelerators SDK comes with a Visual Profiler [2] as well.

Research based performance tools have embraced hardware accelerators as well. Most work, however, focuses on GPU computing using CUDA or OpenCL. The tools typically fall into two categories: existing HPC tools that extend to include hardware accelerators or new tools designed to cope with specific properties of a single accelerator. Examples of the former are TAU [10], the integrated performance monitor (IPM) [5], and the HPCToolkit [11]. A common feature of these three tools is their profiling background. While TAU and HPCToolkit use the callbacks from the NVIDIA CUPTI API to acquire GPU related performance data, IPM uses library wrapping to intercept GPU-host interaction on the API level. GPU Ocelot [4] is a very powerful tool from the second category. While it is primarily a code development framework, it also allows code instrumentation for parallel thread execution (PTX) programs and, thus, can be used to gather performance data as well. Like the vendor tools, however, it lacks the ability to extend to all levels of parallelism.

8.3 Gathering Accelerator Related Performance Information

The combination of trace-based performance analysis with hardware accelerators has to find answers to three major challenges: (1) Dealing with asynchronous accelerator APIs, (2) timing and tracking of evens in a different execution domain, and (3) gathering useful additional information. This section explains the nature of

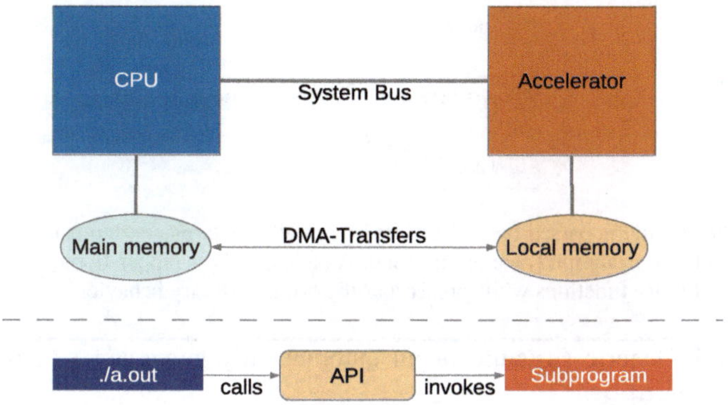

Fig. 8.1 Typical layout of a system using hardware accelerators: it has two disjoint memory locations and the program execution on the accelerator is host driven

these challenges and how they can be overcome to provide an event logging infrastructure capable of accurately recording accelerator usage. It, furthermore, highlights NVIDIA CUPTI as an example how cooperation between well established HPC tool developers and vendors can provide a fertile tool infrastructure adapted to one specific accelerator.

8.3.1 Accelerator APIs

All current HPC related hardware accelerators follow the same principle: A host program running on the CPU that also runs the operating system connects to the available accelerator(s) via an API provided by a system library, usually as part of the accelerator driver or SDK, to offload compute work onto the accelerator device (see Fig. 8.1). The latest FPGA based acceleration environment from Convey is an exception to this scheme since the FPGA can also initiate work to be carried out by the CPU. Other commonalities between the accelerators are that they break up work into parallel blocks executed in SIMD or SPMD fashion and that there is usually no direct means to synchronize or communicate between the blocks since their execution order is not predetermined. Furthermore, the accelerators use an own instruction set architecture (ISA) with an own compiler but allow for program code that combines parts to be executed on the accelerator and on the host. The special compiler separates the two parts, generates an accelerator binary for the accelerator, and passes the host part to a host compiler. One notable exception to this paradigm is OpenCL, which only embeds the source code for the accelerator in the host binary and uses just in time compilation by the accelerator driver at run time.

The *IBM Cell/B.E. processor* is in itself a special kind of hardware accelerator. It bears close resemblance to the combined CPU/GPU processors from AMD (called APUs) and NVIDIA (e.g. Tegra) but was developed earlier. While it also

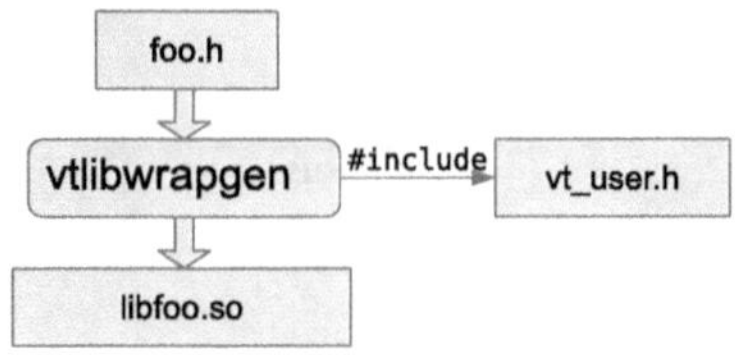

Fig. 8.2 A header file is parsed by `vtlibwrapgen`, all functions are duplicated, and enter and leave statements from VampirTrace are inserted. As a result the wrapper library records entering and leaving of library functions while preserving the original library behavior

Algorithm 1 Generic structure of an automatically generated wrapper function using `vtlibwrapgen`

```
foo_orig ← DLSYM(foo)
function FOO(arg1)
    RECORD_FUNCTION_ENTER(foo,arg1)
    result ← FOO_ORIG(arg1)
    RECORD_FUNCTION_LEAVE(foo)
    return result
end function
```

contains a (quite small) CPU—the POWER processing element (PPE)—its accelerator devices (Synergetic processing elements or SPEs) are much loosely coupled than for example Streaming Multiprocessors in an NVIDIA GPU, SIMD engines in an AMD GPU or a Clearspeed processor. The SPEs can run arbitrary programs, communicate via memory transactions and access the same main memory as the PPE. The Cell/B.E. software ecosystem also shares features described earlier: The SPEs are based on a different ISA than the PPE, hence they require a special compiler whose output is embedded in the host program running on the PPE to launch work on the SPEs during its run time. Previous work [6] has shown that tracing can be extended to also track execution of hybrid applications on the IBM Cell/B.E. processor.

Recording Accelerator API Invocations A logical first choice to record hardware accelerator usage is to extend the measurement infrastructure to also capture accelerator API invocations. This can be achieved without modifying the library to be observed, as shown using static relinking in [15]. This approach, however, is I/O focused and lacks the flexibility to address any kind of programming interface. As proposed in my previous work such limitations can be overcome by a flexible dynamic library wrapping infrastructure [3]. In its workflow (see Fig. 8.2) a header file (e.g. `cuda.h`) is parsed and a wrapper library is generated. Each call in the wrapper library has the same generic structure as shown in Algorithm 1. As a result, VampirTrace can now bind against an unmodified binary program (using `vtrun`) and record the usage of the wrapped library by this program. When applied to hardware accelerator APIs, this method will generate statistics on how the host is interfacing with the device driver. It does, unfortunately, not cope very well with the largely asynchronous nature of accelerator APIs as shown in the following section.

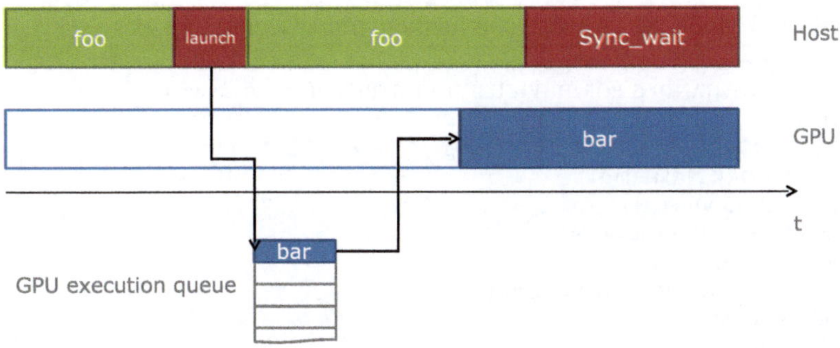

Fig. 8.3 Execution of code on a hardware accelerator is asynchronously launched from the controlling host process, i.e. the host has no influence when the accelerator will begin executing the offloaded work

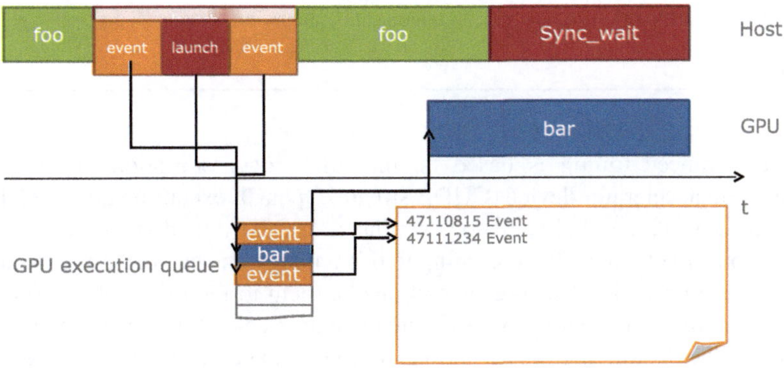

Fig. 8.4 Timing events can be introduced on the device by the wrapper library before and after the offloaded work. This will generate time stamps for the beginning and the end of the program executed on the device

8.3.2 *Tracking Accelerator Events*

Most hardware accelerators simply use execution queues to have work assigned to them by the controlling host. Just tracing the execution of the host application and its interaction with the accelerator API is insufficient to draw conclusions as to how the accelerator is utilized [3, 7] which can be easily seen in Fig. 8.3. In order to allow the user to time the offloaded work, many accelerator APIs offer means to generate time stamps on the accelerator. The Cell/B.E. SPE for example offers the call `spu_read_decrementer()`, CUDA `CudaEvents`, and OpenCL `clGetEventProfilingInfo(...)`. The wrapper library for the accelerator API as introduced in Sect. 8.3.1 can be extended to also time accelerator activity by modifying the respective `launch` command to also record timing information for the launched accelerator work as shown in Fig. 8.4 and Algorithm 2.

Algorithm 2 Modified `launch` command to time accelerator work

```
launch_orig ← DLSYM(launch)
device ← GET_HANDLE_FOR_DEVICE_EVENT_STREAM(device_context)
function LAUNCH(arg1)
    RECORD_FUNCTION_ENTER(launch,arg1)
    PUSH_WORK_NAME(arg1)
    ENQUEUE_EVENT(t1)
    result ← LAUCH_ORIG(arg1)
    ENQUEUE_EVENT(t2)
    RECORD_FUNCTION_LEAVE(launch)
    return result
end function
function SOME_SYNCHRONOUS_OPERATION
    ...
    t1,t2 ← READ_EVENTS(t1,t2)
    name ← POP_WORK_NAME
    RECORD_DEVICE_FUNCTION_ENTER(name,t1,device)
    RECORD_DEVICE_FUNCTION_LEAVE(name,t2,device)
    ...
end function
```

This advanced timing is based on the following assumption: the execution stream of an accelerator device (CUDA stream, OpenCL execution queue, Cell SPE program) is in its parallel semantics similar to a CPU thread spawned by a CPU process to offload work [7]. According to this analogy one can reuse the features of VampirTrace to track CPU threads and invoke them to record accelerator activity. One major difference between CPU threads and accelerator execution streams is that due to the disjoint memory locations for the host and the device the latter do not share the same main memory as CPU threads would. The accelerator programming paradigms call for explicit data transfers between the two memory locations. They are, in this regard, very similar to MPI programming where explicit transfers copy data from one memory location over some kind of network—in the case of hardware accelerators this would be PCI-Express—to another location. Hence, it is easy to note that this transfer can be logged just like an MPI message would be. In order to separate the two kinds of messages in a multi-hybrid program, accelerator transfers are recorded using a special flag so that they can be easily (de-)selected during later analysis. In the same manner accelerator threads are also placed into two special process groups during tracing: The accelerator group contains all accelerator event streams and the accelerator communication group contains all accelerator event streams and the governing host processes.

Another challenge when using accelerator-based time stamps—as in the introduced approach—is timer synchronization. This can, however, be solved rather easily. If one surrounds a synchronous device interaction (e.g., waiting for work completion or a synchronous data transfer) with device events as well, one will receive correlated time stamps on the host and the device, since the call of the accelerator API on the host also generates time stamps on the host on entering and leaving the routine. By using these correlated time stamps it is possible to (1) translate device time stamps into host time stamps and (2) adjust the clock drift of the two non-synchronized clocks during program execution. Since the Cell/B.E. offers a synchronized clock signal over all PPEs/SPEs one only has to synchronize all local clocks once at the beginning and does not have to compensate for clock drift on this accelerator platform.

8.3.3 Including Accelerator Specific Data

On top of the pure timing information for hardware accelerated code segments, it is of interest how well the offloaded program code actually runs. Two main performance questions arise: (1) Is my program execution actually accelerated by the use of the accelerator device(s) and (2) How close to peak performance is my accelerator program? To answer these questions the performance tool has to acquire information from the hardware accelerator while it is executing the program. This can be done by modifying the accelerator program to generate performance information, which is impractical for a number of reasons: for one, most accelerator compilers do not offer an instrumentation interface like CPU compilers. Thus, for trace-based performance analysis of offloaded routines one has to inject performance gathering routines into the source code of the device program either manually or by source-to-source transformations. The means to do so are accelerator specific and lack some generality that would offer an analogy to the library wrapping for accelerator API interception. Furthermore, recording performance data directly on the device requires additional resources to store the data. This can lead to very large run time modifications due to the measurement, e.g. due to registers spilling into local memory. In the worst case the accelerated program does not fit any longer onto the accelerator. This is especially true on FPGAs where the recompilation would also potentially too time consuming.

A second, more practical, option is to record additional device information, e.g. hardware performance counters. These counters can also identify good or bad performance and accelerator utilization. The main challenge is to access them since, unlike on the CPU and on the Cell/B.E., they are usually not directly accessible from the accelerator device. Thus, a performance tool is depending on the accelerator vendor exposing access to the performance counters via an API. Fortunately, for all major hardware accelerators such an API is now available (Cell: Cell Performance Counter tool, NVIDIA: CUPTI, AMD: GPUPerfAPI).

The performance counters are usually sampled with registered callbacks from the accelerator library. This means that a typical usage scheme (with one launch and one synchronization call as shown in Fig. 8.5) will only provide two counter samples that, as shown before, do not necessarily align with the beginning and end of a accelerated program. Currently the only way to overcome this limitation is to use an additional host thread to frequently poll the performance counters.

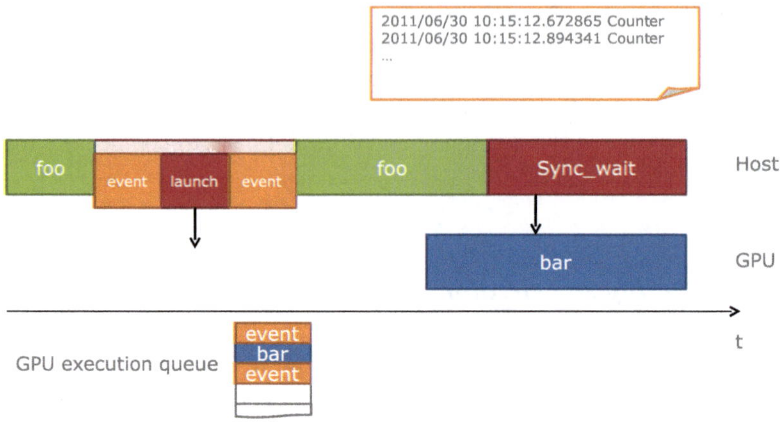

Fig. 8.5 VampirTrace can register callbacks into all CUDA functions to poll performance counter information whenever a CUDA function is invoked. The quality of the counter values does, however, highly depend on the frequency of CUDA calls

8.4 Example: Integration of CUDA and OpenCL into VampirTrace/Vampir

VampirTrace and Vampir are well established performance tools for recording and visualizing, respectively, traces of massively parallel applications [12]. As previously discussed the performance data recording tool VampirTrace has been extended in multiple ways to handle hardware accelerators. While prior work included a special version of VampirTrace for the Cell/B.E., the library wrapping approach presented in this paper has been used to cover CUDA and OpenCL. The automatically generated wrapper libraries have been extended to record device activity in its own event stream and to record data transfers between the host and the device as messages. Furthermore, on NVIDIA devices access to hardware performance counters is available using the CUDA performance tool interface (CUPTI). VampirTrace will detect a CUDA or OpenCL installation automatically during its `configure` run, if they are installed in the default location. It will then compile the provided CUDA or OpenCL wrapper libraries that include all previously described enhancements.

The usage of the GPUs can be traced in two different ways: (1) without recompiling the application or (2) by recompiling the application using the VampirTrace compiler wrapper to also instrument the host code. In the former case one has to use the `vtrun` utility to execute the application, e.g. by invoking

```
mpirun -np 4 vtrun ./a.out
```

where `a.out` is an MPI application also using CUDA or OpenCL. The `vtrun` command uses `LD_PRELOAD` to bind the VampirTrace library against the application, thus, intercepting all calls to the MPI, pthread, CUDA, and OpenCL libraries.

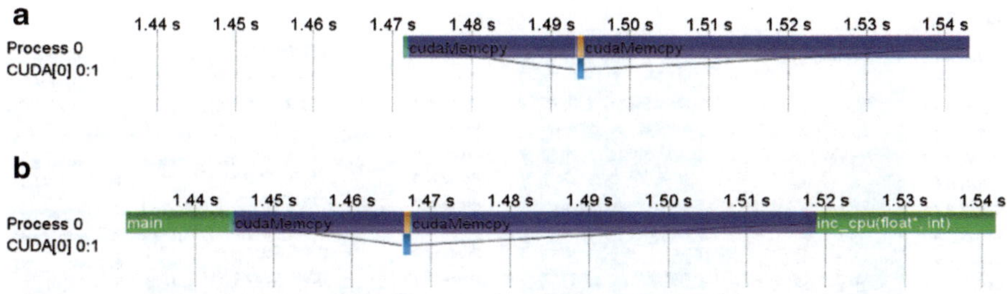

Fig. 8.6 VampirTrace can track usage of CUDA devices by just intercepting the CUDA library calls (**a**) or by also instrumenting the host code (**b**)

Hence, one will not receive any information about activity on the host and only see how the processes exchange MPI messages and how they interface with the GPUs. The GPU activity, however, is complete since all GPU actions are invoked through the corresponding CUDA or OpenCL library. The timeline of a very simple CUDA application invoking just one kernel is shown in Fig. 8.6a.

When also instrumenting the host code, one receives a full picture of the applications behavior on all levels of parallelism. In contrast to the previous example, Fig. 8.6b also shows detailed host activity. An example of a performance study of a fully hybrid application using MPI, pthreads, and CUDA simultaneously, as shown in Fig. 8.7, illustrates how the holistic tracing approach of VampirTrace offers an unrivaled level of detail. The detailed time stamps for the entire program activity enable the analysis of load balance on the parallelism levels individually but also concurrently, so that scalability bottlenecks can be identified. The statistics displays available within Vampir always produce their content based on the current time interval selected in the pipeline and the selected event streams. Thus, the best use of the statistics can be accomplished by looking at the parallelism levels individually.

It would be helpful, if Vampir were able to distinguish between processes, threads, and accelerators. Vampir is limited by the capabilities of the underlying trace format OTF [9] which treats all event streams the same regardless of their origin. The trace format does, however, allow a parent definition for an event stream so that a process-thread-mapping can be accomplished. In order to keep this behavior—and with it the backwards compatibility of OTF—no new means to distinguish different kind of children (threads or accelerators) were introduced. Instead, accelerator event streams are treated the same as host threads and their ID is chosen by selecting the next available thread ID for the associated host process. Additionally, the accelerator name and its ID is included in the event stream label. Vampir will always draw child event streams directly below their parent so that the program hierarchy is visible in the Master Timeline.

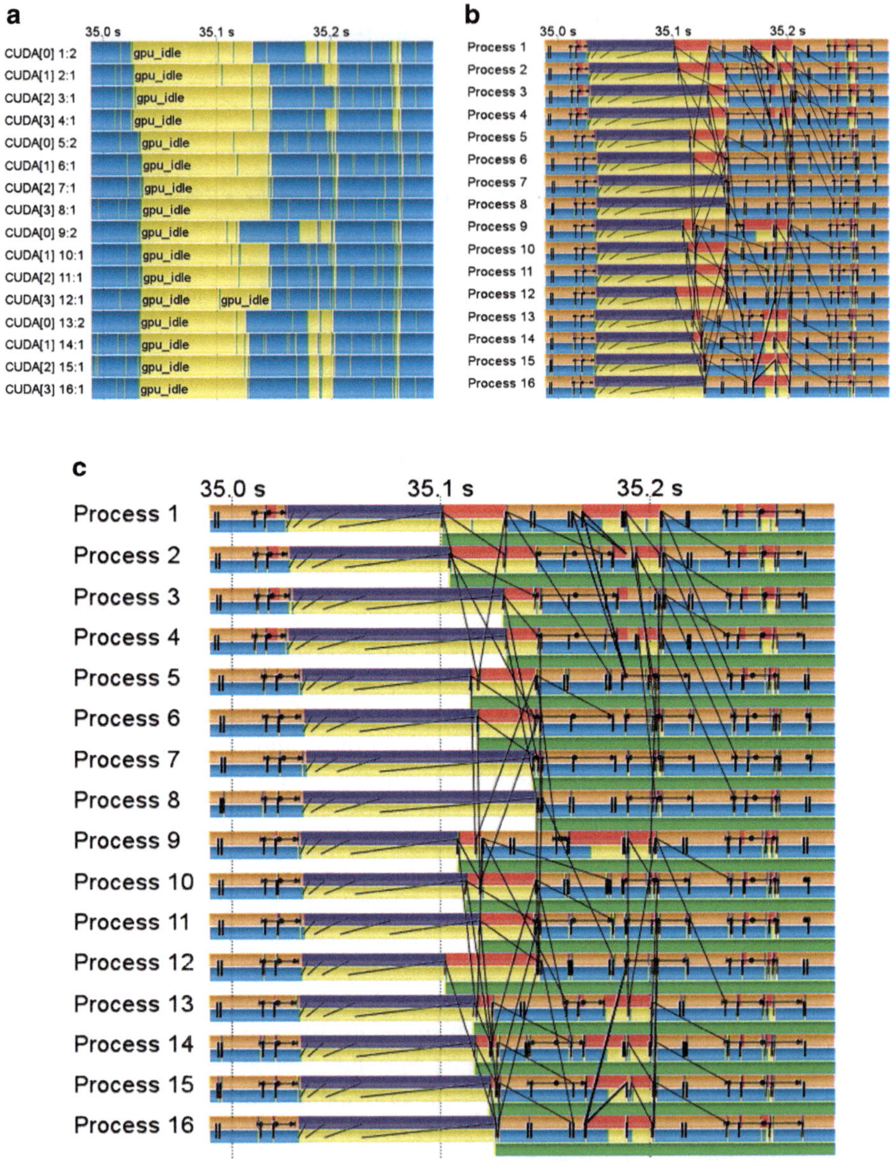

Fig. 8.7 The analysis of a multi-hybrid application can select different parallelism levels for visualization. In (**a**) only the CUDA devices are shown during a phase where all GPUs are idle. By also showing the activity of the controlling host processes in (**b**) one can see, that the devices are idle due to large cudaMemcpys. In (**c**), where all participating processes, threads, and CUDA streams are shown concurrently, it becomes clear that this is done prior to a host thread starting its work. In this case all data is pulled from the GPU to start a host thread for writing the data to disk for post mortem video generation

8.5 Summary and Future Work

This paper presented how event logging/tracing can be extended to hardware accelerators. It demonstrated that trace infrastructures like VampirTrace that can already handle hybrid applications can be extended to also record accelerator usage. It has to be noted that this is accomplished by re-utilizing already established metrics (threads and messages) since it is not desirable to change the fixed trace format which would influence the whole analysis tool chain. The paper also presented how CUDA and OpenCL tracing is included in VampirTrace and how this affects visualization with Vampir. A result of the presented work is that modifications to the trace format are not really necessary to deal with the additional level of parallelism introduced by the usage of hardware accelerators. Their performance data can well be recorded and visualized as is. Nevertheless, the sheer amount of trace data will overload the current time line visualization. Novel approaches to present very large amounts the performance data are required to help the performance analyst in locating e.ġ. load imbalance in parallel applications. Future work also includes extending VampirTrace to also work with AMD's GPUPerfAPI.

Acknowledgements The research work reported here was supported by hardware grants from NVIDIA as part of a CUDA Research Center and AMD as well as UT Batelle subcontract 4000096585.

References

1. AMD: APP Profiler. Online. http://developer.amd.com/tools/AMDAPPProfiler (2011)
2. Clearspeed: Visual Profiler. Online. http://www.clearspeed.com/products/sdk_details.php (2011)
3. Dietrich, R., Ilsche, T., Juckeland, G.: Non-intrusive performance analysis of parallel hardware accelerated applications on hybrid architectures. In: International Conference on Parallel Processing Workshops, San Diego, pp. 135–143 (2010). doi: 10.1109/icppw.2010.30
4. Farooqui, N., Kerr, A., Diamos, G., Yalamanchili, S., Schwan, K.: A framework for dynamically instrumenting gpu compute applications within gpu ocelot. In: Proceedings of the Fourth Workshop on General Purpose Processing on Graphics Processing Units, GPGPU-4, pp. 9:1–9:9. ACM, New York, NY (2011). doi: http://doi.acm.org/10.1145/1964179.1964192. http://doi.acm.org/10.1145/1964179.1964192
5. Fuerlinger, K., Wright, N.J., Skinner, D.: Comprehensive performance monitoring for gpu cluster systems. In: Proceedings of the 12th IEEE International Workshop on Parallel and Distributed Scientific and Engineering Computing (PDSEC), in conjunction with IPDPS-11, Anchorage, AK (2011). http://projekt17.pub.lab.nm.ifi.lmu.de/fuerling/research/pubs//FUERLINGER_2011_PDSEC.pdf
6. Hackenberg, D., Brunst, H., Nagel, W.E.: Event tracing and visualization for cell broadband engine systems. In: Luque, E., Margalef, T., Benítez, D. (eds.) Euro-Par 2008, LNCS 5168, pp. 172–181. Springer, Berlin/New York (2008). doi: 10.1007/978-3-540-85451-7_19
7. Hackenberg, D., Juckeland, G., Brunst, H.: Performance analysis of multi-level parallelism: inter-node, intra-node and hardware accelerators. Concurr. Comput. Pract. Exp. (2011). doi: 10.1002/cpe.1725

8. IBM: PA-SDK3-Tool. Online. http://www.ibm.com/developerworks/power/tutorials/pa-sdk3tool (2011)
9. Knüpfer, A., Brendel, R., Brunst, H., Mix, H., Nagel, W.E.: Introducing the open trace format (otf). In: Alxandrov, V.N., Albada, G.D., Sloot, P.M.A., Dongarra, J.J. (eds.) 6th International Conference on Computational Science (ICCS), vol. 2, pp. 526–533. Springer, Reading (2006)
10. Mayanglambam, S., Malony, A., Sottile, M.: Performance measurement of applications with GPU acceleration using CUDA. In: International Conference on Parallel Computing (PARCO 2009), Lyon (2009)
11. Mellor-Crummey, J., Tallent, N., Liu, X.: Hpctoolkit: new capabilities, ongoing work, and challenges ahead. In: CScADS Summer 2011 Workshops – Performance Tools for Extreme Scale Computing, Tahoe City, CA (2011). http://cscads.rice.edu/workshops/summer-2011/slides/performance-tools/Rice-HPCToolkit-CScADS-2011.pdf
12. Müller, M.S., Knüpfer, A., Jurenz, M., Lieber, M., Brunst, H., Mix, H., Nagel, W.E.: Developing scalable applications with vampir, vampirserver and vampirtrace. In: Bischof, C., Bücker, M., Gibbon, P., Joubert, G.R., Lippert, T., Mohr, B., Peters, F.J. (eds.) Parallel Computing: Architectures, Algorithms and Applications. Advances in Parallel Computing, vol. 15, pp. 637–644. IOS Press, Amsterdam (2008). http://www.booksonline.iospress.nl/Content/View.aspx?piid=8455
13. NVIDIA: CUDA Visual Profiler. Online. http://developer.download.nvidia.com/compute/DevZone/docs/html/C/doc/Compute_Visual_Profiler_User_Guide.pdf (2011)
14. NVIDIA: Parallel NSight. Online. http://developer.nvidia.com/nvidia-parallel-nsight (2011)
15. Vijayakumar, K., Mueller, F., Ma, X., Roth, P.C.: Scalable I/O tracing and analysis. Technical report, Department of Computer Science, North Carolina State University; Computer Science and Mathematics Division, Oak Ridge National Laboratory (2009)

Chapter 9
Folding: Detailed Analysis with Coarse Sampling

Harald Servat, Germán Llort, Judit Giménez, Kevin Huck, and Jesús Labarta

Abstract Performance analysis tools help the application users to find bottlenecks that prevent the application to run at full speed in current supercomputers. The level of detail and the accuracy of the performance tools are crucial to completely depict the nature of the bottlenecks. The details exposed do not only depend on the nature of the tools (profile-based or trace-based) but also on the mechanism on which they rely (instrumentation or sampling) to gather information.

In this paper we present a mechanism called folding that combines both instrumentation and sampling for trace-based performance analysis tools. The folding mechanism takes advantage of long execution runs and low frequency sampling to finely detail the evolution of the user code with minimal overhead on the application. The reports provided by the folding mechanism are extremely useful to understand the behavior of a region of code at a very low level. We also present a practical study we have done in a in-production scenario with the folding mechanism and show that the results of the folding resembles to high frequency sampling.

9.1 Introduction

Application users are typically delighted when they are granted access to a new supercomputer because they expect their applications to run at a faster pace than before. Although that each new supercomputer reports faster results than their predecessors, it is unquestionable that user applications only reach a portion of the peak performance of the supercomputer.

H. Servat (✉) · G. Llort · J. Giménez · K. Huck · J. Labarta
Barcelona Supercomputing Center, Universitat Politècnica de Catalunya, c/Jordi Girona, 31, 08034 Barcelona, Catalunya, Spain
e-mail: harald.servat@bsc.es; german.llort@bsc.es; judit.gimenez@bsc.es; kevin.huck@bsc.es; jesus.labarta@bsc.es

H. Brunst et al. (eds.), *Tools for High Performance Computing 2011*, 105
DOI 10.1007/978-3-642-31476-6_9,

Performance analysis tools aim to explain to the user the reasons why his or her application cannot reach the supercomputer's peak performance to eventually optimize the application and increase its performance. Nowadays we can classify performance tools by the amount of data gathered at runtime: profile-based and trace-based. While profile-based tools keep timeless performance data of the execution run by aggregating the collected data, trace-based tools generate a sequence of timestamped events that report the progression of the application. In addition, performance tools can use two different methods to gather performance metrics sampling and instrumentation. One the one hand, sampling relies on periodic signaling to gather performance metrics, on the other hand, instrumentation injects code to specific locations of an application to emit the metrics. To increase the details of such methods the user may decrease the period of the sampling or instrument additional code locations, but with the extra overhead drawback. An alternative is to combine both instrumentation and sampling as in Extrae [4], TAU [15] or Scalasca [19].

In this paper we describe folding [13, 14], a mechanism for trace-based performance tools, that provides high level of detail using Paraver [12] traces. The folding mechanism combines sampled and instrumented information gathered from the sample application in order to produce more detailed results than its original form in the Paraver tracefile. Moreover, folding benefits from long running applications because it can be combined with high periodicity sampling to produce detailed results with low overhead cost. We have combined the folding mechanism with a density-based clustering tool [5, 6] to develop a framework that automatically provides characteristics of the most time-consuming computation regions in an application. This allows the analyst to focus on potential performance issues on specific code locations that represent the relevant part of the application.

The rest of this paper is structured as: Sect. 9.2 reviews both sampling and folding concepts and describes how the performance counters and the callstack information is folded. Section 9.3 shows the folding results for a set of analyzed applications. We compare the quality of the folding results with fine grained sampling results in Sect. 9.4. Finally, we discuss related and future work in Sects. 9.5 and 9.6, respectively.

9.2 Folding: Instrumentation and Sampling

We extended the Extrae [4] instrumentation package with sampling capabilities. The sampling mechanism can either rely on the PAPI middleware (`PAPI_overflow`) or the regular operating system alarms. The sampling handler is responsible for gathering hardware performance counters and a segment of the whole callstack at the sample point. We depict how the overhead increases the execution time of the application when using smaller sampling periods in Fig. 9.1. Naturally, the more samples the library gathers the more the application becomes perturbed due to the intrusion of the library. For example, when the sampling period is $50 * 10^3$

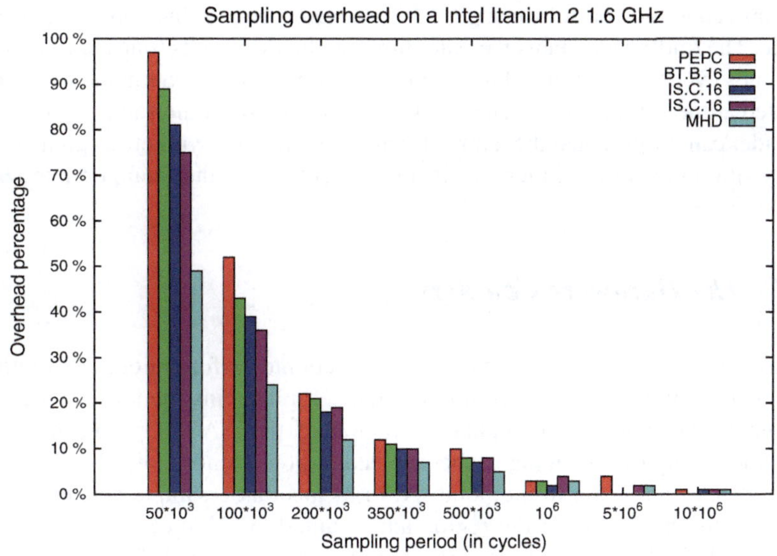

Fig. 9.1 Overhead of the sampling mechanism at different sampling periods when using different MPI applications on a Intel Itanium2 supercomputer

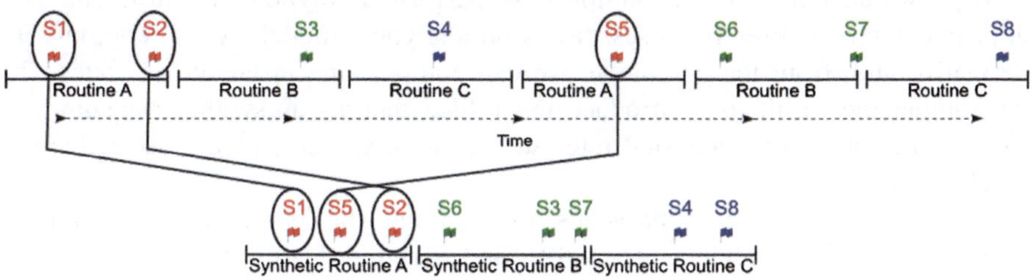

Fig. 9.2 Example of the folding mechanism. The *top* timeline shows the samples belonging to the different routines. In the *bottom* figure, we plot how the folding maps the samples on the different synthetic regions. We have outlined the mapping for the *Routine A* which involves the samples labeled as *S1, S2* and *S5*

cycles the application can take up to 100 % more time to run. This large overhead makes the fine grained sampling not desirable because of the large impact it has on the application.

The folding mechanism takes benefit of long running applications (which is very common in the scientific computing) to combine sampled information from different regions into a single synthetic region. This mechanism combines instrumented and sampled information to increase the details of the instrumented regions. Within the folding process, each instrumented and sampled information plays a role. While instrumented information is used to delimit regions, sampled information determines how the performance behavior evolves within the region it belongs. Consider the time-line shown in Fig. 9.2 where an iterative application executes three instrumented routines named A, B, C. Each single flag represents a taken sample. The folding mechanism creates three synthetic regions (one per routine) and then maps every sample into those regions according to the routine that was being executed at the sample point. The result is shown below in the same Figure, where the reader can observe that the mapped samples within the synthetic region preserve their position in respect of their original region instead of their temporal ordering.

9.2.1 The Hardware Counters

The folding results, regarding the hardware counter information, are plotted in Fig. 9.3. This plot shows the evolution of the completed instructions counter using the folding samples in a computation region of the NAS BT benchmark [11]. The folded samples belonging to the selected region are normalized and shown as red crosses. A red cross in the point (X,Y) means that a sample was taken at the X % from the beginning of the region and counted the Y % of the total metric in the region. In both figures, we can infer temporal phases with different performance characteristics also noting their sequence and duration by solely using the folded samples.

Once we have the cloud of samples, we perform a polynomial adjustment. The adjustment serves three purposes: (a) as an analytical model we can compute its derivative and, from the derivative, we can compute the instantaneous rates, (b) we sample the result to reintroduce the folded metrics as synthetic events into the tracefile at a user requested rate, and (c) it serves also as a noise reduction mechanism.

We explored several approaches to compute the polynomial adjustment: polynomial fitting, Bézier curves [2] and Kriging [18] interpolation. Polynomial fitting requires choosing the grade of a polynomial which cannot be done independently from the data. In addition, choosing a low order polynomial will give soft but inaccurate fitting, while a high order polynomial will fit better but will result in big fluctuations. Bézier curves do not require additional data but the points themselves also fitted well on our tests except on the stationary points. Finally, the Kriging interpolation, which is a general version of the Bézier curve, works with the sample points plus some interpolation parameters (including fitting strictness). After some tests, we found a typical combination of parameters that fitted the samples well even in stationary points.

In Fig. 9.3, the contouring results and the instantaneous rate are shown in the plot as a green and blue lines, respectively. Now comparing the information derived from the adjustment and its derivative, we can easily identify four different phases by their MIPS rate. Each phase involves a period in which the routine runs really fast (at more than 6,000 MIPS) and follows a period where the routine runs slower (at 2,000 MIPS, approximately).

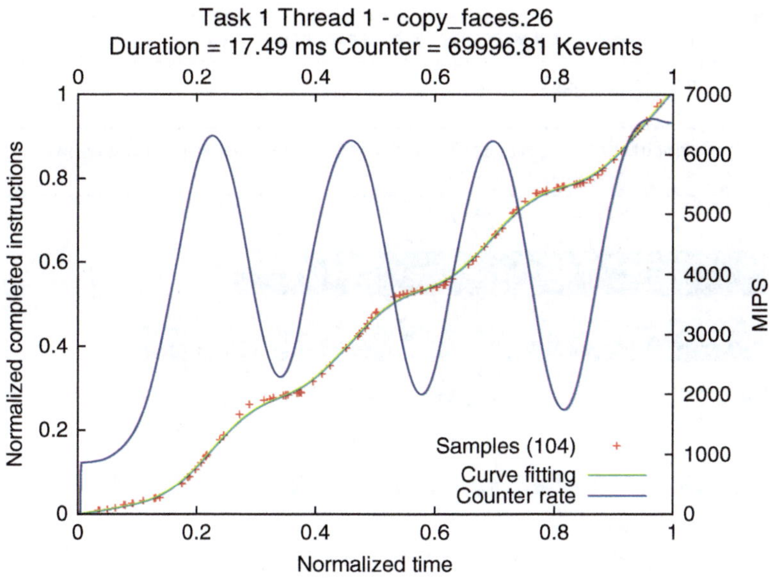

Fig. 9.3 Folding results of the instructions counter for a computation region in the NAS BT benchmark when executed in a SGI Altix 4700 machine

9.2.2 *The Callstack*

The folding mechanism is also applied to the sampled callstack information if the application contained debugging information during the execution. In contrast to the folded hardware counters, folding the callstack information does not benefit from an analytical model like Kriging. The folded callstacks are emitted to the tracefile and can be used to display which code region was being executed at a certain time in the synthetic regions. Although this method can be sensitive to noise, it provides an approximate location where to start looking in the application code. With such information, a time-based analysis tool like Paraver can create correlations between the folded callstack information and any metric. Such correlation can subsequently stored and then display those metrics in GVIM (a *vim*[1] editor using a *GNOME*[2] front-end) using an add-on we built using the GVIM scripting mechanism. We show a screenshot of this add-on on Fig. 9.4. In this Figure, the lines that achieved the lowest and the highest MIPS are annotated in colors, ranging from a minimum in green to a maximum in blue.

[1]http://www.vim.org

[2]http://www.gnome.org

```
c-------------------------------------------------------------------
c      loop over all cells owned by this node
c-------------------------------------------------------------------
       do c = 1, ncells

c-------------------------------------------------------------------
c      compute the reciprocal of density, and the kinetic energy,
c      and the speed of sound.
c-------------------------------------------------------------------
          do k = -1, cell_size(3,c)
             do j = -1, cell_size(2,c)
                do i = -1, cell_size(1,c)
                   rho_inv = 1.0d0/u(1,i,j,k,c)
                   rho_i(i,j,k,c) = rho_inv
                   us(i,j,k,c) = u(2,i,j,k,c) * rho_inv
                   vs(i,j,k,c) = u(3,i,j,k,c) * rho_inv
                   ws(i,j,k,c) = u(4,i,j,k,c) * rho_inv
                   square(i,j,k,c)     = 0.5d0* (
     >                 u(2,i,j,k,c)*u(2,i,j,k,c) +
     >                 u(3,i,j,k,c)*u(3,i,j,k,c) +
     >                 u(4,i,j,k,c)*u(4,i,j,k,c) ) * rho_inv
                   qs(i,j,k,c) = square(i,j,k,c) * rho_inv
                enddo
             enddo
          enddo
```

Fig. 9.4 GVIM capture showing the MIPS within the editor according to the stats captured at runtime and the results of the folding mechanism

9.3 Example of Usage

To exemplify the usage, we will show the study we did of the Code_Saturne [3] application, which also belongs to the PRACE Benchmark Suite [16]. The application was executed on MareNostrum, a 10,240 core supercomputer based on two dual-core PowerPC 2.3 GHz processors per blade interconnected with a Myrinet network. The application was compiled with the IBM XL Fortran compiler version 12.1 using `-O3 -qstrict` and the experiment ran 200 time-steps using 32 cores with a sampling period of 100 ms and MPI instrumentation.

We used a clustering tool [5, 6] that characterizes computation regions according to their metrics characteristics to find the application structure. The results for the characterization are shown in Fig. 9.5. Each dot within the scatterplot represents a computation region (i.e., a region within an MPI exit point and the consecutive MPI entry point). We classified every computation region by two performance metrics of the region: its executed instructions and its IPC (instructions per cycle). This classification shows on one hand the total amount of work done by a portion of code and the speed achieved to execute that amount of work. After the classification, the computation regions are grouped (colored in the plot) together by their distance to each other using a density-based algorithm and the grouping information is sent back to the tracefile for further analyses. From all the resulting groups we focused on the more relevant ones. Concretely, we will focus on those computation regions that last more than 50 ms, and from them we will focus on Cluster 1 because

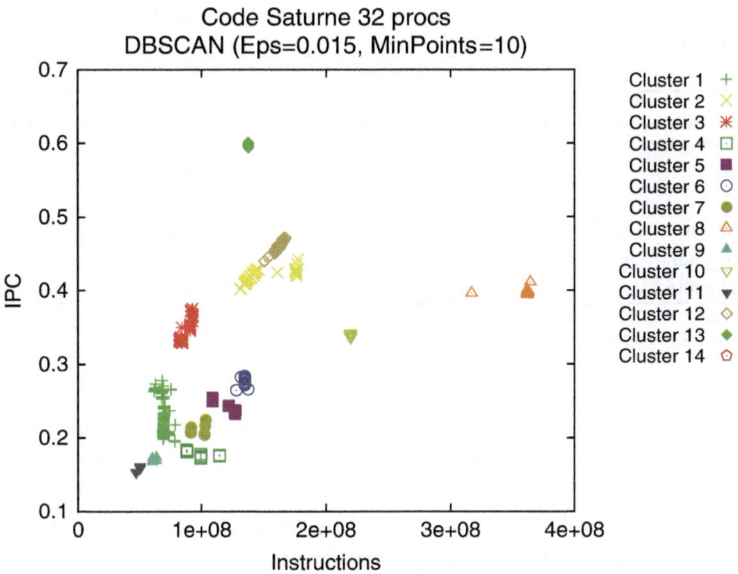

Fig. 9.5 Scatterplot shows the grouping of the different computation regions of the Code_Saturne application. The Y-axis shows the measured IPC whereas the X-axis the committed instructions

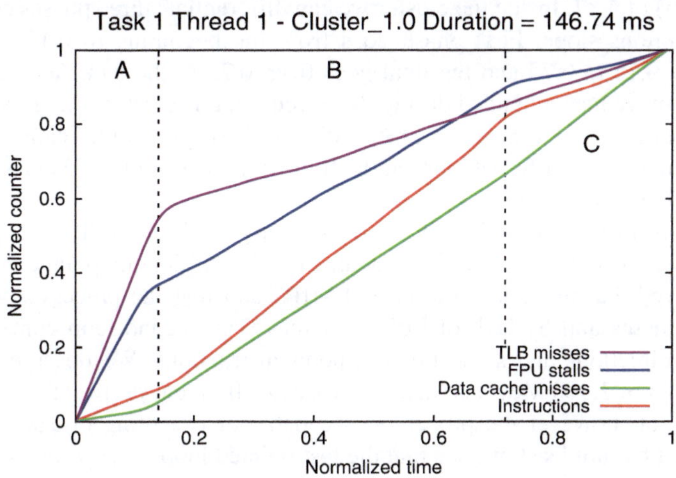

Fig. 9.6 Plot depicting the evolution of several performance counters (namely: instructions, data cache misses, FPU stalls and TLB misses) within Cluster 1 in the Code-Saturne application

it has more than 20K occurrences and accounts more than the 35 % of the total computation time.

To study the internal behavior of Cluster 1, we applied the folding mechanism on these regions. Due to space restrictions, we have summarized the results we obtained in Fig. 9.6. In this figure we show the evolution of four different performance

Table 9.1 Average values for different metrics in Code_Saturne code divided by phases using the slope of the instructions slope

Computation region	Cluster 1		
Phase name	A	B	C
Location in `gradrc.f90`	751–753	786–802	972–985
Duration[a]	20.54	84.82	41.08
MIPS	300	600	300–450
Data cache stall cycles[b]	400	1,700	2,000
TLB miss stall cycles[b]	160	20	20
FPU stall cycles[b]	550	160	75

[a]In milliseconds
[b]In millions per second

counters (instructions, data cache misses, FPU stalls and TLB misses) within Cluster 1 using the interpolation results of the folded samples. We have manually divided the plot by the slope of the instructions counter rate to divide different computation phases (labeled as A, B and C in this Figure) within the same computation region. In Table 9.1 we show different metrics and the code location for each phase.

According to the metrics we extracted from the Paraver tracefile we obtained that `Cluster 1` ran at approximately 472 MIPS which is very bad for the MareNostrum processor because it represents that it achieved a CPI (cycles per instruction) of 4.87. Inside the cluster we can differentiate three phases considering the instructions slope. First phase goes from the beginning to 0.15, the second goes from 0.15 to 0.75 and the final goes from 0.75 to the end. According to the information we have obtained during the execution this cluster is mainly executing the routine contained in `gradrc.f90`. After analyzing the folded callstack, we can relate them to three different loops at the following lines at lines 751–753, 786–802 and 972–985, respectively.

The loop in lines 751–753 shows a CPI between 7 and 9. This loop accesses seven different vectors and performs a large number of floating point computations. We observed that this phase ran at 300 MIPS and that the processor was stalled by TLB misses and by lack of FPU resources. The second loop contains lots of pointer indirections and some floating point instructions. We observe that 1,700 Mcycles out of 2,300 Mcycles (i.e., the processor frequency) are stalled due to data cache misses, however the processor is capable of executing instructions at 600 MIPS. Last but not least, we see that the last pointed loop has a worse performance behavior in terms of stalled cycles due to cache misses resulting in a worse MIPS rate (ranging from 300 to 450).

We unsuccessfully tried to improve the performance of the application. We splitted the first loop into three loops to prevent the cache to hold all the vectors referenced, however after applying this modification the average duration of Cluster 1 increased by 5 ms. Changing the second loop would need changing the face numbering algorithm to increase the locality of the accessed data and/or rethinking this part of the code to reduce the number of indirections. Also, it would

Table 9.2 Experiment setup for the quality study

	Application execution setup	
Application name	BT.B	
Sampling mode	Detailed	Coarse
Sampling period	50 Kcycles	10 Mcycles
Samples per second	32,000	160
Sampling overhead	89 %	2 %
Number of iterations	1	200
Total number of samples per task	5,661	5,445

be interesting to reduce the number of accesses within the loop because there are currently 11 arrays accessed. Finally, we applied blocking and splitting techniques to the third loop to increase the locality of the data accesses. However, when applying blocking with a block size of 128 elements, it made Cluster 1 3 % slower whereas splitting the loop made Cluster 1 a 20 % slower, mainly due to the increase of the TLB misses.

9.4 Validation of the Results

We validated and studied the quality of the folding mechanism by using the BT.B benchmark from the NAS MPI Parallel Benchmark Suite 3.2 on a SGI Altix machine with Intel Itanium 2 processors running at 1.6 GHz. Although this benchmark is heavily optimized, some computing regions of BT.B show a non-uniform behavior (as seen in Fig. 9.3) making it a nice candidate to study. The Kriging contouring algorithm used in the folding mechanism behaves as a low-pass filter, so with the higher number of samples being interpolated the interpolation results are more detailed. Because of this, the experiments are focused on validating how the folded and interpolated results resemble finer sampled metrics. Details regarding the experimentation are shown in Table 9.2.

The first experiment is intended to compare the shape of the hardware counter metrics in high frequency sampled traces and low frequency folded sampled traces. We did the comparison with two of the most time consuming computation regions from the benchmark. To facilitate the reading, each computation region is named according to the routine to which it belongs. We compare one time-step of the high frequency trace to the folded results of all iterations. The comparison is done using different counters available on the chosen processor, namely committed instructions, executed floating point operations and branches, and L2 cache accesses, hits, and misses.

The Kriging implementation is limited to return a vector of equidistant points which prevents us to get the value of the interpolation at any arbitrary point. To perform the study, we compared the results between the samples of the reference

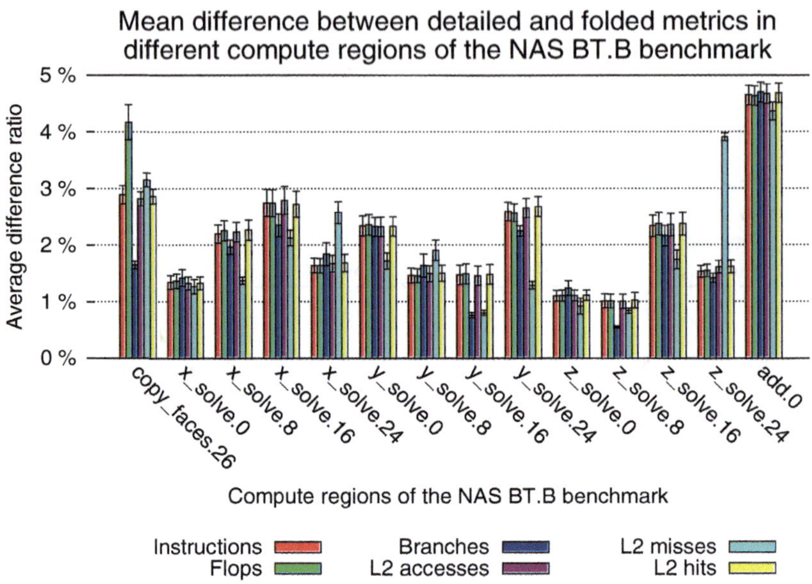

Fig. 9.7 Study of absolute mean difference ratios and square mean errors on different applications comparing high detail sampling and folded results on the NAS BT.B benchmark

time-step and the folded samples by interpolating both of them in the same number of points and then we compute the absolute mean difference and the square mean error by comparing every resulting slice.

Results of the comparison are shown in Fig. 9.7. The X-axis on the plot depicts the representative computation regions in the NAS BT benchmark, whereas the Y-axis represents the percentage of variation between the detailed sampling and the folded results using coarse sampling and the square mean error by the bars and the whiskers respectively. In the plot we observe mean differences up to 5 % for every pair performance counter and computation region. The coarse-grained sampling generates approximately the same number of samples for the whole run as the reference iteration because of the number of the time-steps iterated on the coarse-grain execution is the ratio between the two sampling frequencies. So we conclude that using scattered samples on long execution runs we can obtain a good approximation of the performance metrics without harming the gathered performance metrics due to the sampling overhead.

In the second experiment we ran measured how the number of folded samples influences the absolute mean difference shown in the previous experiment. To carry out the comparison we took as a reference the highly detailed sampled information of the longest computation regions of the BT.B benchmark from the previous experiment, and then we computed the mean difference varying the number of samples folded on the coarse-grained execution.

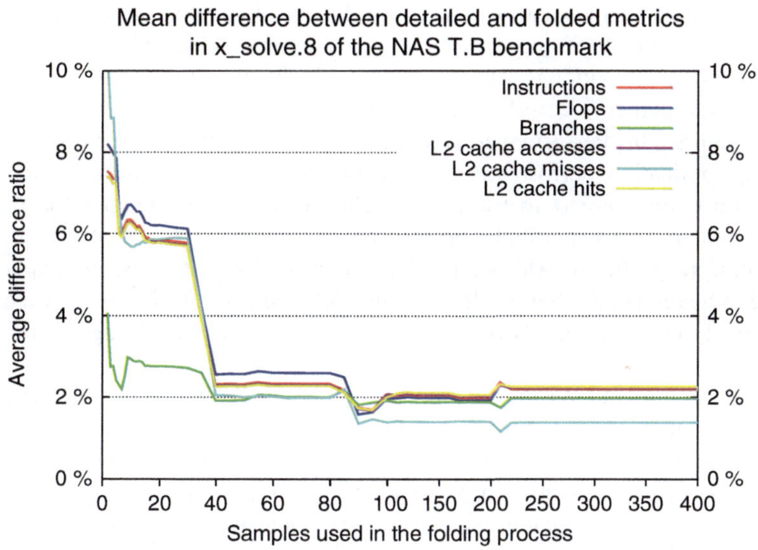

Fig. 9.8 Evolution of the mean difference between highly detailed sampling and folded results in a BT.B compute region of the x_solve routine

The plot in Fig. 9.8 shows that the absolute mean difference varies depending on the number of samples folded in the region. This plot shows that the mean absolute difference decreases as the number of samples increase. Precisely, the number of samples on which the difference is below 5 % is about 40 for the computation region we present and using more samples just leads to marginal differences. So we can conclude that there is no need for extremely long executions to get similar results between folding coarse grain samples and using detailed sampling, although this may vary on the application.

9.5 Related Work

The gprof [7] profiler uses sampling and instrumentation to gather function call counts and estimate the time spent per function. This profiler requires the application to be compiled and linked with a special flag that instructs the compiler adding monitoring to count the number of calls to user routines, while it uses sampling to accumulate time to the user routines. This tool provides summaries for simple metrics, whereas the folding mechanism combines both sampling and instrumentation providing much more details resulting in more helpful analysis.

HPCToolkit [17] uses sampling with trampoline optimizations to provide call-stack details. HPCToolkit can also show sampled callstack data in a time-line using a recently added *hpctraceview*. Our mechanism also provides instrumentation and sampling data in a Paraver time-line but is also capable of combining the performance data within the tracefile to provide highly detailed synthetic regions. The synthetic information is reintroduced into the Paraver tracefile so all the metrics can be used together to correlate them easily.

The Sun Studio Performance Analyzer [8] is a set of tools for collecting and viewing application performance data using tracing and profiling mechanisms. These tools are able to instrument synchronization calls, heap allocation and deallocation routines, parallel regions and constructs, and can also sample the application to profile it. Although being a powerful set of tools, the experiments needed to be repeated if the results were not detailed enough. The work we propose is designed to produce highly detailed performance metrics avoiding repetitive data collection.

Azimi, Stumm and Wisniewski are the authors of [1] where they present an on-line performance analysis tool that gathers counter values periodically. This tool displays the evolution among the selected counters in a time-line breakdowns the cycles-per-instruction using a model called Statistical Stall Breakdown. Although they offer some instrumentation capabilities, their work is just focused on the sampling mechanism to characterize the whole system instead of applications.

The TAU framework has recently added sampling capabilities in its measurement system [10]. And, although they gather performance counter information, their work is mainly focused on instrumentation cooperating with the sampling to augment the TAU profiles with PC callstack information instead of increasing the details of the performance counters. The cooperation to increase their profiles is based on creating keys to the application callstack and to the TAU event stack during the application execution to reduce the information presented to the user in order to produce a summarized view.

Finally, the Scalasca tool has also been extended using sampling. As in TAU, their developers put the major effort to provide information about application routines instead of providing performance counters. Scalasca provides accurate time metrics by subtracting the time in MPI calls of an user routine for a sampling period when PMPI instrumentation is turned on. Also in Scalasca, as in HPCToolkit, trampolines are used to replace the returning addresses in the callstack to reduce the overhead in the unwinding process.

9.6 Conclusions and Future Directions

We have evaluated the quality of the results of using low frequency sampling and applying the folding mechanism compared with high frequency sampled data. The results show good resemblance (less than 5 % of difference) between them, even using only a fraction of the samples of the whole run. It makes low frequency sampling and folding a good alternative to high frequency sampling without penalizing the application. We have shown an example of utilization of the folding when combined with the clustering. Yet the example we provided was not improved, we have shown successful studies and optimizations in [14].

Future work could include working on a better correlation between performance counters at this finer detail to thoroughly describe the performance behavior and also to locate hot and cold spots within the code. Another interesting future topic would be to perform an on-line study of the application using the work described in [9]. The combination of these two mechanism would provide detailed performance information as the application executes, showing the performance at different application stages and stop when enough samples have been gathered and study whether the sampling periodicity matches any period of the computation bursts.

Acknowledgements This work is granted by the IBM/BSC MareIncognito project and by the *Comisión Interministerial de Ciencia y Tecnología* (CICYT) under Contract No. TIN2007-60625.

References

1. Azimi, R., et al.: Online performance analysis by statistical sampling of microprocessor performance counters. In: ICS '05: Proceedings of the 19th Annual International Conference on Supercomputing, pp. 101–110. ACM, New York (2005). doi:http://doi.acm.org/10.1145/1088149.1088163
2. Bézier, P.: Numerical Control. Mathematics and Applications. Wiley, London (1972). Translated by: A.R. Forrest and Anne F. Pakhurst
3. Code Saturne. http://research.edf.com/research-and-the-scientific-community/softwares/code-saturne/introduction-code-saturne-80058.html. Accessed July 2011
4. Extrae Instrumentation Package. http://www.bsc.es/paraver. Accessed August 2012
5. González, J., et al.: Automatic detection of parallel applications computation phases. In: IPDPS'09: 23rd IEEE International Parallel and Distributed Processing Symposium, Rome, Italy. IEEE Computer Society, Piscataway (2009)
6. González, J., et al.: Automatic evaluation of the computation structure of parallel applications. In: PDCAT '09: Proceedings of the 10th International Conference on Parallel and Distributed Computing, Applications and Technologies, Hiroshima, Japan. IEEE Computer Society, Hiroshima (2009)
7. Graham, S.L., et al.: Gprof: a call graph execution profiler. In: SIGPLAN '82: Proceedings of the 1982 SIGPLAN Symposium on Compiler Construction, pp. 120–126. ACM, New York (1982). doi:http://doi.acm.org/10.1145/800230.806987
8. Itzkowitz, M.: Sun studio performance analyzer. http://developers.sun.com/sunstudio/overview/topics/analyzer_index.html. Accessed August 2012
9. Llort, G., et al.: On-line detection of large-scale parallel application's structure. In: IEEE International Symposium on Parallel and Distributed Processing (IPDPS), 19–23 April 2010, pp. 1–10. doi: 10.1109/IPDPS.2010.5470350. URL:http://ieeexplore.ieee.org/stamp/stamp.jsp?tp=&arnumber=5470350&isnumber=5470342 (2010)
10. Morris, A., et al.: Design and implementation of a hybrid performance measurement and sampling system. In: ICPP 2010: Proceedings of the 2010 International Conference on Parallel Processing, San Diego, California (2010)
11. NAS Parallel Benchmark Suite. http://www.nas.nasa.gov/Resources/Software/npb.html. Accessed August 2012
12. Pillet, V., et al.: Paraver: a tool to visualize and analyze parallel code. In: Nixon, P. (ed.) Transputer and occam Developments, pp. 17–32. IOS Press, Amsterdam (1995). http://www.bsc.es/paraver. Accessed July 2011

13. Servat, H., et al.: Detailed performance analysis using coarse grain sampling. In: Euro-Par Workshops (Workshop on Productivity and Performance, PROPER), Delft, The Netherlands pp. 185–198. Springer Berlin, Heidelberg (2009)
14. Servat, H., et al.: Unveiling internal evolution of parallel application computation phases. In: ICPP'11: International Conference on Parallel Processing, Taipei, Taiwan (2011)
15. Shende, S.S., Malony, A.D.: The TAU parallel performance system. Int. J. High Perform. Comput. Appl. **20**(2), 287–311 (2006). doi:http://dx.doi.org/10.1177/1094342006064482
16. Simpson, A.D., Bull, M., Hill, J.: Identification and categorisation of applications and initial benchmarks suite (2008). http://www.prace-project.eu/documents/Identification_and_Categorisatio_of_Applications_and_Initial_Benchmark_Suite_final.pdf. Accessed July 2011
17. Tallent, N., et al.: Hpctoolkit: performance tools for scientific computing. J. Phys. Conf. Ser. **125**(1), 012088 (2008)
18. Trochu, F.: A contouring program based on dual Kriging interpolation. Eng. Comput. **9**(3), 160–177 (1993)
19. Wolf, F., et al.: Usage of the SCALASCA for scalable performance analysis of large-scale parallel applications. In: Tools for High Performance Computing, pp. 157–167. Springer, Berlin/Heidelberg (2008)

Chapter 10
Advances in the TAU Performance System

Allen Malony, Sameer Shende, Wyatt Spear, Chee Wai Lee, and Scott Biersdorff

Abstract Evolution and growth of parallel systems requires continued advances in the tools to measure, characterize, and understand parallel performance. Five recent developments in the TAU Performance System are reported. First, an update is given on support for heterogeneous systems with GPUs. Second, event-based sampling is being integrated in TAU to add new capabilities for performance observation. New wrapping technology has been incorporated in TAU's instrumentation harness, increasing observation scope. The fourth advance is in the area of performance visualization. Lastly, we discuss our work in Eclipse Parallel Tools Platform.

10.1 Introduction

The TAU performance system [12] has been in development and use in the high-performance computing (HPC) community for over 20 years. During this time, it was the changes in the parallel architectures, systems, software development, and applications that make up HPC technologies that generated requirements for continual improvement in the TAU toolset and the methodology it supports. Today's HPC evolution is no different in the demands it places on next-generation performance tools. The benefit of an established system like TAU and others, is in the ability to leverage existing capabilities to create new and more powerful features. Of course, it can also be more challenging if advances force re-engineering of the toolset's architecture and implementation. The following paper highlights recent advances in the TAU performance system in five areas that cover different aspects of the changing needs of HPC. First, we discuss updates on our work to integrate into TAU support for measurement of heterogeneous systems using GPUs. Second,

A. Malony (✉) · S. Shende · W. Spear · C.W. Lee · S. Biersdorff
University of Oregon, Eugene, OR, USA
e-mail: malony@cs.uoregon.edu; sameer@cs.uoregon.edu; wspear@cs.uoregon.edu; cheelee@cs.uoregon.edu; scottb@cs.uoregon.edu

H. Brunst et al. (eds.), *Tools for High Performance Computing 2011*,
DOI 10.1007/978-3-642-31476-6_10,

we describe the addition of event-based sampling in TAU to address limitations of a purely probe-based approach for performance observation. Instrumentation has always been an important technology for enabling performance measurement. The third area covers new wrapping technology that has been incorporated in TAU's instrumentation harness. The need to makes sense of high-dimensionality performance data motivates better methods for data analysis and presentation. The fourth advance we will discuss is in the area of performance visualization. Lastly, we discuss our work in Eclipse Parallel Tools Platform.

10.2 Instrumentation Of GPU Accelerated Code

Understanding the performance of scalable heterogeneous parallel systems and applications will depend on addressing new challenges of instrumentation, measurement, and analysis of heterogeneous components, and integrating performance perspectives in a unified manner. Here we will cover an approach to addressing these requirements undertaken by the TAU, PAPI [3] and Vampir-Trace [4] development teams. We will examine the approach in terms of computation and measurement models in order to ground the discussion on tool implementation. The measurement techniques supported by the tools are intended to be practical solutions for these approaches with respect to present technology.

10.2.1 Synchronous Method

The validity of the time measurement is predicated on when the kernel issued to the accelerator begins execution and when it ends. We use the term synchronous to indicate that it is the CPU (host) who is observing the begin and end events, as denoted by the diamonds in Fig. 10.1. Measurements are made on the CPU by recording events before kernel launch and after synchronization. If the host immediately waits on kernel termination after launch, its kernel measurement is, in effect, synchronized with the kernel execution. In this case, the measurement method is equivalent to measuring a subroutine call. In essence, a synchronous approach assumes the kernel will execute immediately, and the interval of time between the begin and end events will accurately reflect kernel performance. Unfortunately, this assumption is overly restrictive and leads to inaccuracies when more flexible modes of kernel execution are used. As the figure suggests, the host need not block after kernel launch and it can be a long time before it is synchronized with the kernel, resulting in poor estimates of actual kernel execution time. Moreover, multiple kernels can be launched into a stream or multiple streams before a synchronization point is encountered. The benefit of a synchronous approach is that it does not require any additional performance measurement mechanisms beyond what is presently available.

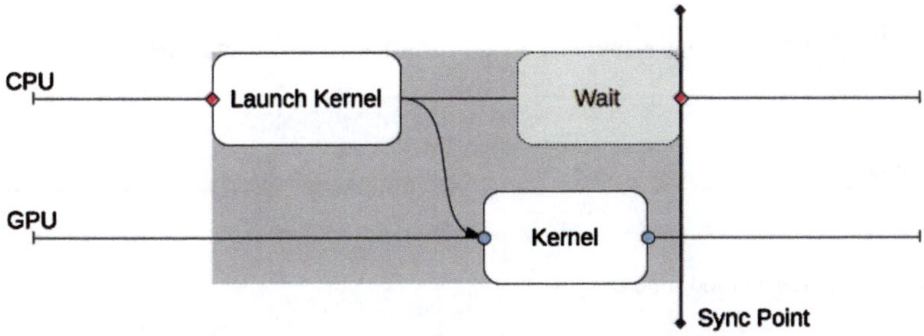

Fig. 10.1 Synchronous method timeline. *Shaded area* represents the execution time

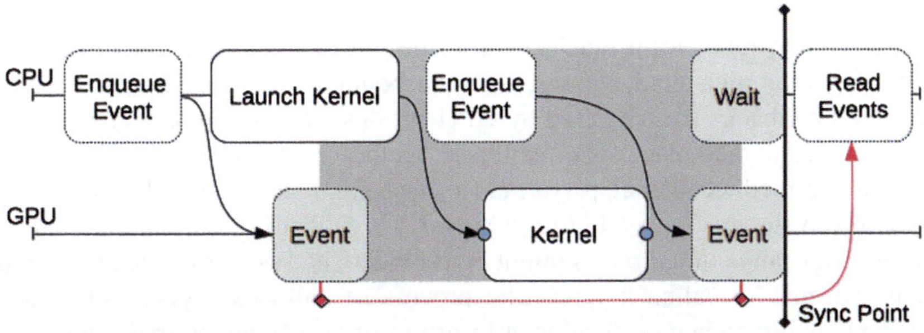

Fig. 10.2 Event queue method timeline. *Shaded area* represents the measured execution time on the host device

10.2.2 Event Queue Method

The main problem with the synchronous approach is that the kernel execution is measured indirectly, not by the GPU. Consider a special type of kernel called an event kernel which will record the state of the GPU when it is executed. If we could inject an event kernel into the stream immediately before and after the computational kernel, it would be possible to obtain performance data more closely linked with kernel begin and end. While it is the responsibility of the host to generate the event kernels, queue them into the stream, and read the results, it is the underlying GPU device layer that will take responsibility for making the measurement. Measurements are made on the event kernels placed in the same stream as the computational kernel. In theory this method, shown in Fig. 10.2, works well. It adequately addresses the case where multiple kernels are launched in a stream, if each is wrapped by an associated event kernel they are all accounted for even if the synchronization point is not until much later. However, there are a few practical downsides. First, it relies entirely on the device manufacture to provide support for the event kernel concept. The notion of events is a part of both the CUDA and OpenCL specification. However, restrictions on how events can be used and what performance information is returned is implementation dependent.

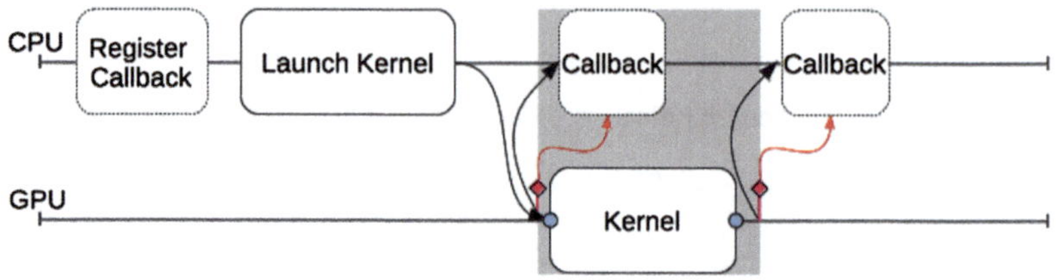

Fig. 10.3 Callback method timeline

10.2.3 Callback Method

A third method relies on a mechanism in the device layer that triggers callbacks on the host for registered actions, like the begin and end of kernel execution. Registered callbacks are triggered by GPU actions, allowing more tightly coupled measurements to take place. The "callback" method portrayed in Fig. 10.3 suggests that more immediate kernel performance measurement is possible since control can be given directly to a CPU process via the callback. It is also more flexible since a wider range of callbacks might be provided, and performance measurement can be specific to callback type. The process of callback registration makes it possible to avoid code modification at locations of kernel launch in the application. Clearly, a callback method is dependent on the device manufacturer to provide support in the device layer and even the GPU hardware. The research presented here demonstrates support for GPU performance measurement with CUDA and OpenCL in three well-known performance tools PAPI, VampirTrace, and the TAU Performance System. The tools targeted in this paper support performance counter measurement, profiling, and tracing for scalable parallel applications, and are generally representative of probe-based measurement systems.

10.2.4 TAU Performance System Implementation

TAU [12] has tools for source instrumentation, compiler instrumentation, and library wrapping that allows CPU events to be easily observed. In particular, they allow library wrapping of the CUDA runtime/driver API and preloading of the wrapped library prior to execution. Then, each call made to a runtime or driver routine is intercepted by TAU for measurement before/after calling the actual CUDA routine. TAU library interposition happens dynamically with the Linux LD_PRELOAD mechanism and can be used on an un-instrumented executable. Such features are commonly used in other performance tools. For instance, VampirTrace also applies the LD_PRELOAD mechanism with the CUDA libraries [5]. Clock synchronization is needed to correct any time lag between the CPU and the GPU, just as when tracing events across multiple nodes. This is accomplished by measuring simultaneously (or as nearly as possible) at a synchronization point the time on CPU and the time on the GPU. The GPU time can be obtained at the synchronization event which has just occurred. By measuring the difference between these two times we can measure the clock lag between the CPU and GPU.

10.3 Event Based Sampling (EBS)

Our integration of the TAU performance system with event-based sampling measurement, called TAUebs, represents our approach to a hybrid measurement system which includes both probe-based and sampling-based components. A hybrid measurement system attempts to marry the strengths of both probe- and sampling-based measurements while mitigating their weaknesses.

Sampling-based measurements determine an application's performance by statistical observation via some interrupt mechanism. Probe-based measurement systems instruments the application code at specific points to collect performance data at the time the instrumented code is executed. Morris et. al. [10] detailed the design and implementation of a prototype for such a hybrid measurement system. The initial prototype focused on the generation of sampling traces followed by a post-mortem merge operation with probe-based TAU profiles generated from the same application run. Building on this prototype, we have now implemented support for tighter integration of sampling-based information into TAU's profile data structures at application runtime. The tool employs the same sampling technology based on work by HPCToolkit [1], Perfsuite [11] and PAPI [3].

When a TAU-instrumented application encounters an interrupt to take a sample, TAU's event path data structures are consulted to discover the current TAU event context. A basic information-gathering approach is simply to capture the number of times a program counter value is encountered. A histogram of sample counts against encountered program counter values is maintained for each unique TAU event path context at runtime. At the end of the application, before TAU profiles are written to disk, TAU's event paths are augmented by performance information derived from sampling-based measurements as follows:

1. An intermediate event representing the sum of all sampled events that were encountered in the context of some TAU event path P is created as a new leaf node to path P.
2. For each program counter value found in the histogram maintained for path P, attempt to resolve the source file name, function name and line number. Multiple program counter values mapping to the same line number in the code are treated as the same sample event. Metric values are then assigned by multiplying the known sample period with the number of samples. Finally, each sample event are added to TAU's event path as leaf nodes.

The sample information collected for each TAU event path is flat within the context of that TAU event path. This design is illustrated in Fig. 10.4. We are

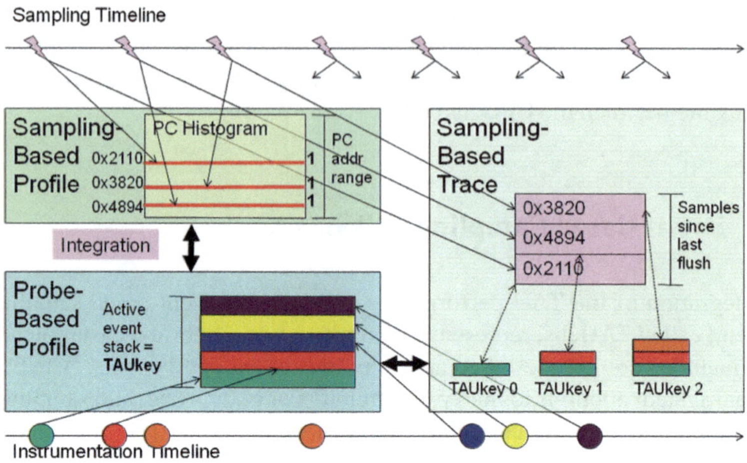

Fig. 10.4 An illustration of the complementary sampling and instrumentation methodologies. Instrumentation provides sampling context via an event stack

working toward incorporating stack unwinding for the runtime integration of sample information to TAU profiles. We are also working on resolving issues associated with the correct mapping of program counters to code structure in the face of code optimization. We have implemented several solutions to mitigate issues associated with asynchronous signal safety and thread safety. In particular, we did not wish to pre-allocate memory for entire code address spaces for program counter histograms. As a result, we had to make use of pool storage allocators in C++ to reduce the chances of making our own memory allocation calls when our signal handle happened to interrupt some memory allocation event in application code.

Szebenyi et. al. [15] presented results on a similar hybrid measurement effort developed on top of the Scalasca set of tools. They specifically insert probe-based measurements on MPI calls while capturing the performance of the rest of the application through sampling-based measurements. They have described their approach for addressing some of the common issues we both face in our efforts.

10.4 Automatic Wrapper Library Generation

Many parallel applications are constructed using software library packages with interfaces callable from standard programming languages. Packages are often layered, internally calling other libraries to implement underlying functionality, which can be hidden to the user. Having an ability to intercept package calls at library routine interfaces enables performance tools to gather both semantic (contextual) and performance data for analysis purposes. Shende et al. [13] discusses this wrapper feature in detail in the context of tracking IO events. Below is a summary of this feature.

TAU can automate the creation of wrapper libraries using its `tau_gen_wrapper` tool. It parses the interface of a library (as represented in a header file) and creates an wrapped interface for each routine, as may be specified in a selective instrumentation file. This interface starts and stops TAU timers around the original function call. There are three ways to generate the wrapper library:

1. Redefining the function using a pre-processor macro
2. Preloading the wrapper library at runtime
3. Using linker-based substitution of a routine with an alternative implementation.

This tool supports all three cases. It internally uses the Program Database Toolkit (PDT) to parse the header file.

This can be done by defining a header file that internally redefines the name of a routine as a macro that redirects all references to the given call with another. The compiler's pre-processor then replaces all references to the original call at the callsite in the source code with the corresponding call defined by the tool (e.g., read replaced by tau_read).

The above approach works well for C and C++ programs where library calls are replaced explicitly during compilation. Unfortunately, this approach does not extend well to Fortran programs. Moreover, the instrumentation technique is limited to application code regions where the source code is available for recompiling.

This method also relies explicit re-compilation but includes support for Fortran. TAUs instrumentation tool (`tau_instrumentor`) examines the source code, its PDB file as generated by the Program Database Toolkit (PDT), and re-writes the Fortran calls in the instrumented source code. In this method, performance measurement code is inserted directly in the source code.

Many HPC operating systems such as Linux, Cray Compute Node Linux (CNL), IBM BlueGene Compute Node Kernel (CNK), Solaris permit pre-loading of a library in the address space of an executing application specifying a dynamic shared object (DSO) in an environment variable (LD_PRELOAD). It is possible to create a tool based on this technique that can intercept library operations by means of a wrapper-library where the all calls are redefined to call the global routine (identified using the dlsym system call) internally. This method only supports dynamic executables—IBM BlueGene and Cray XE6 and XK6 systems default to static executables but dynamic executables can be requested.

10.5 3D Visualization

It has always been challenging to create new performance visualizations, for three reasons. First, it requires a design process that integrates properties of the performance data (as understood by the user) with the graphical aspects for good visual form. This is not easy, if one wants effective outcomes. Second, unlike visualization of physical phenomena, performance information does not have a natural semantic visual basis. It could utilize a variety of graphical forms and visualization types (e.g., statistical, informational, physical, abstract). Third, with increasing application concurrency, performance visualization must deal with the problem of scale. The use of interactive three-dimensional (3D) graphics clearly helps, but the visualization design challenge is still present.

In addition to these challenges, there are also practical considerations. Because of the richness of parallel performance information and the different relationships to the underlying application semantics, it is unreasonable to expect just a few performance visualizations to satisfy all needs. Where visualization of large performance information does exist, it is generally embedded as "canned" displays in a profile or trace analysis tool. If a user has a different concept in mind, they have very limited ability to make changes.

To move towards a more general method of creating 3D visualizations, we considered the salient components of the existing versions and the need to specify aspects of a 3D design in a more flexible manner. Two ideas resulted: (1) separate the visualization layout design from visualization user interface (UI) design, and (2) allow the properties of each to be specified by the user. We have added these capabilities to the ParaProf [2] profile analysis tool.

Visualization layout design is concerned with how the visualization will appear. Our approach allows the visual presentation to be specified with respect to the parallel profile data model (events, metrics, metadata) and possible analysis of this information [14]. Two basic layout approaches we support are mapping to Cartesian coordinates provided by MPI and filling a space of user-defined dimensions in order of MPI rank. We have also worked to develop a specification language for describing more complex layouts of thread performance in a 3D space. In our initial implementation of these custom layouts, mathematical formulae define the coordinates and color value of each thread in the layout. The formulae are based on variables provided by the profile data model. These input variables include event and metric values for the current thread being processed as well as global values such as the total number of threads in the profile. The specification is applied successively to each thread in the profile to determine X, Y and Z coordinate values and color values which are used to generate the visualization graphics. Our initial implementation for expression analysis uses the MESP expression parser library [9]. MESP provides a simple syntax for expressing mathematical formulae but is powerful enough to allow visualization layouts based on architecturally relevant geometries or the mathematical relationship of multiple performance variables.

The layout of points in a visualization which conforms to a useful pattern, reflecting the physical layout of the system or the data decomposition of the application, may be densely packed or otherwise obscured. Therefore it is also necessary to provide the user with a means to selectively display salient features of a performance topology. For example, as shown in Fig. 10.5, we can exclude the points with middling values and display only the outliers. In addition to clarifying the structure of performance behavior with respect to machine topology, this also helps to highlight potential topologically sensitive performance problems such as oversubscribed or underutilized nodes.

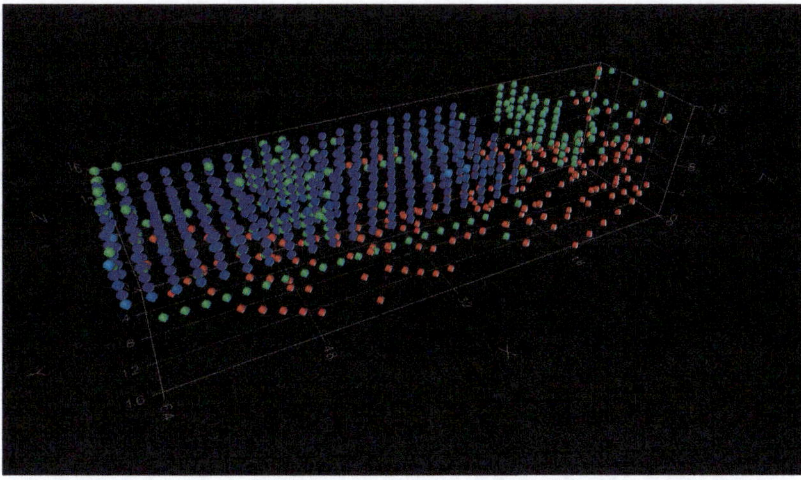

Fig. 10.5 16k-core 3D topology map of Sweep function in the Sweep 3d application on BG/L. Only nodes with high (*red*) and low (*blue or green*) values are shown

Visualization UI design is concerned with how the visualization will be controlled. The key insight here is to have the UI play a role in "binding" data model variables used in the layout specification. This approach implements the functionality present in the current ParaProf views, where the user is free to select events and metrics to be applied in the visualization as inputs to layout formulae. However, for large performance proles of many threads/processes, the specied layout can result in a dense visualization that obscures internal structures. The current ability to zoom and rotate the topology in the UI partially ameliorates this issue. Our model for visualization UI further allows more sophisticated ltering techniques.

10.6 Eclipse Integration

Tau has supported integration with the Eclipse IDE [6] for some time now. We wrap the standard command-line based TAU analysis workflow in the graphical Eclipse interface. This simplifies the process of performance analysis, ties it more closely with the overall development cycle in Eclipse and exposes the capabilities of the TAU performance system through the UI. By integrating with the Eclipse Parallel Tools Platform (PTP) [7] we also take advantage of new IDE-based parallel and HPC development capabilities.

During the development of the TAU plug-ins it became evident that much of the work being done was applicable to other performance analysis systems and similar command-line based tools. At a high level, such tools typically operate on some combination of compilation, execution, and analysis steps and their inputs are similar to those of TAU. To take advantage of this congruity, the workflow logic and User Interface (UI) elements, which were initially hard-coded into the original TAU plug-ins, were converted to a generalized API. Additionally, to make the system more easily accessible and extensible, we developed an XML interface for defining both performance tool workflows and their UIs within Eclipse/PTP. The result is the general-purpose External Tools Framework (ETFw). ETFw allows both tool and application developers to integrate performance analysis systems into an Eclipse environment without the effort and expertise that are required to develop new Eclipse plug-ins. In fact, XML workflow definitions for external performance tools can be added or updated without restarting the Eclipse platform.

Although ETFw generalized much of the hard-coded behavior of the original TAU plug-ins, advanced TAU-specific functionality remains encapsulated within a plugin structure. This functionality includes PAPI hardware counter selection, as shown in the UI example in Fig. 10.6. However, the advanced API extension points used by the TAU-specific plug-ins are available to other tools that require logic or UI elements that are too application-specific for the ETFw to handle.

The ETFw's XML workflow format consists of three fundamental elements, which define the compilation, execution, and analysis steps of the workflow. The order, number, and presence of these steps may vary depending on the intent of the workflow and the employed analysis tools.

- The compilation step assigns compiler commands to be used for the relevant programming languages.
- The execution step defines commands to be composed with the target executable, if any. This covers tools such as Valgrind that take the target application as an input argument.
- The analysis step defines a series of commands that may be run on any data generated during program execution.

Each application or tool defined in an XML workflow may have its command and input parameters specified in the XML file. Alternatively, command-line options may be specified, which will appear in the Eclipse/PTP UI, where the user may enable, disable, and assign values to them dynamically. Once an XML workflow has been composed, it can be modified easily to suit different use cases. It also can be distributed to other users, who can easily load it into their Eclipse/PTP environments and run their applications with the performance analysis workflow, without concerning themselves with tool invocation details. In addition to adding support for arbitrary performance tools, the ETFw's abstraction of performance tool operations simplifies the implementation of more complicated workflows. This includes workflows that require multiple executions of the target application, such as parametric studies [8]. ETFw is now part of the PTP plug-in and is, thus, available to all users with a current Eclipse/PTP IDE configuration.

Formerly TAU and other external tools could only be invoked locally. Much of the power of the PTP lies in its ability to manage development on remote systems. We have extended the ETFw to take advantage of the PTPs remote capabilities. Now a user operating on a local workstation can use the Eclipse environment to build, execute and manipulate performance data from applications on remote,

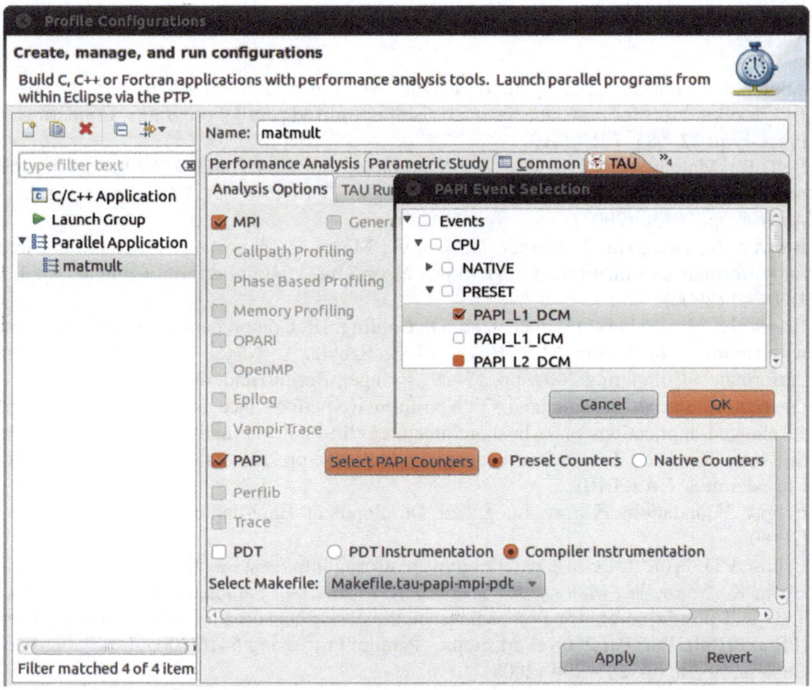

Fig. 10.6 Configuration selection UI for the TAU Eclipse plugin, including popup window for selecting PAPI hardware counters

production machines. This has the potential to make the typical HPC application development cycle much easier and more closely aligned with the standards of conventional software development.

10.7 Conclusion

Parallel performance toolsets must continue to improve to keep abreast of HPC innovations and technology evolution. Five recent advances in the TAU performance system were presented and discussed with respect to various HPC performance measurement and analysis requirements. All of them are or will be part of the TAU open source distribution.

Acknowledgements This research was conducted at the University of Oregon under grants DE-SC0001777, DEFG02-08ER25846, and DE-SC0006723 from the Department of Energy, Office of Science. Resources from NERSC were utilized in this work. This research used resources of the Argonne Leadership Computing Facility at Argonne National Laboratory, which is supported by the Office of Science of the U.S. Department of Energy under contract DE-AC02-06CH11357.

References

1. Adhianto, L., Banerjee, S., Fagan, M., Krentel, M., Marin, G., Mellor-Crummey, J., Tallent, N.: HPCToolkit: tools for performance analysis of optimized parallel programs. Concurr. Comput. Pract. Exp. **22**, 685–701 (2010)
2. Bell, R., Malony, A.D., Shende, S.: A portable, extensible, and scalable tool for parallel performance profile analysis. Proceedings of the EUROPAR 2003 Conference, Klagenfurt, Austria, pp. 17–26 (2003)
3. Browne, S., Dongarra, J., Garner, N., Ho, G., Mucci, P.: A portable programming interface for performance evaluation on modern processors. Int. J. High Perform. Comput. Appl. **14**(3), 189–204 (2000)
4. Brunst, H., Hackenberg, D., Juckeland, G., Rohling, H.: Comprehensive performance tracking with vampir 7. In: Muller, M.S., Resch, M.M., Schulz, A., Nagel, W.E. (eds.) Tools for High Performance Computing 2009, pp. 17–29. Springer, Berlin/Heidelberg (2010)
5. Dietrich, R., Ilsche, T., Juckeland, G.: Non-intrusive performance analysis of parallel hardware accelerated applications on hybrid architectures. In: First International Workshop on Parallel Software Tools and Tool Infrastructures (PSTI 2010), pp. 135–143. IEEE Computer Society, Los Alamitos, CA (2010)
6. Eclipse Foundation: Eclipse Integrated Development Environment. http://www.eclipse.org (2009)
7. Eclipse PTP Project: Eclipse parallel tools platform. http://www.eclipse.org/ptp (2009)
8. Huck, K., Spear, W., Malony, A., Shende, S., Morris, A.: Parametric studies in eclipse with TAU and perfExplorer. In: Proceedings of the Workshop on Productivity and Performance (PROPER08), EuroPar 2008 Workshops – Parallel Processing 5415/2009, Las Palmas de Gran Canaria, Spain, pp. 283–294 (2008)
9. Math Expression String Parser (MESP): http://expression-tree.sourceforge.net/ (2004)
10. Morris, A., Malony, A.D., Shende, S., Huck, K.: Design and implementation of a hybrid parallel performance measurement system. In: International Conference on Parallel Processing, San Diego, pp. 492–501 (2010)
11. National Center for Supercomputing Applications: University of Illinois at Urbana-Champaign, Perfsuite. http://perfsuite.ncsa.uiuc.edu/ (2011)
12. Shende, S., Malony, A.D.: The TAU parallel performance system. Int. J. High Perform. Comput. Appl. **20**(2), 287–311 (2006)
13. Shende, S., Malony, A.D., Spear, W., Schuchardt, K.: Characterizing i/o performance using the tau performance system. In: International Conference on Parallel Proceedings, Parco Exascale Mini-symposium, Ghent, Belgium (2011)
14. Spear, W., Malony, A.D., Lee, C.W., Biersdorff, S., Shende, S.: An approach to creating performance visualizations in a parallel profile analysis tool. In: Workshop on Productivity and Performance (PROPER 2011), Bordeaux, France Aug, 2011
15. Szebenyi, Z., Gamblin, T., Martin, S., de Supinski, B.R., Wolf, F., Wylie, B.J.: Reconciling sampling and direct instrumentation for unintrusive call-path profiling of MPI programs. In: Proceedings of the 25th IEEE International Parallel and Distributed Processing Symposium (IPDPS), May 2011, pp. 637–648. IEEE Computer Society, Anchorage, AK (2011)

Chapter 11
Temanejo: Debugging of Thread-Based Task-Parallel Programs in StarSS

Rainer Keller, Steffen Brinkmann, José Gracia, and Christoph Niethammer

Abstract To make use of manycore processors and even accelerators, several parallel programming paradigms exist, such as OpenMP, CAPS HMPP and the StarSs programming model. All of these programming models provide the means for programmers to express parallelism in the source code, identifying tasks and for all but OpenMP the dependency between those, allowing the compiler and the runtime to schedule tasks onto multiple concurrent executing entities, like threads in a many-core systems. While the programmer may have a good overview of which parts of the code may be run independently as separate tasks on a fine granular level, the overall execution behavior may not be obvious at first. This paper describes the usability features of the newly developed Temanejo debugger.[1]

11.1 Introduction

Parallel programming adds another level of complexity in every respect, especially when debugging the application. Different parallel programming models have different perspectives in that regard – parallel programming models based on processes allow having debuggers attach to each individual process. For programming models that are based on threads, finding bugs is a more difficult feat due nature of threads. In Unix environments, threaded programs share resources such as signals, file

[1]This work was supported by the European Community's Seventh Framework Programme [FP7-INFRASTRUCTURES-2010-2] as project grant "Toward Exaflop Applications" (TEXT) under grant agreement number 261580.

R. Keller (✉) · S. Brinkmann · J. Gracia · C. Niethammer
HLRS, Nobelstrasse 19, Stuttgart, Germany
e-mail: keller@hlrs.de; brinkmann@hlrs.de; gracia@hlrs.de; niethammer@hlrs.de

H. Brunst et al. (eds.), *Tools for High Performance Computing 2011*,
DOI 10.1007/978-3-642-31476-6_11,

```
#include <stdio.h>

#pragma css task input(a) output(out)
void fib(int a, int *out) {
    int tmp1, tmp2;
    if (a <= 1)
        *out = 1;
    else {
        fib (a-1, &tmp1);
        fib (a-2, &tmp2);
        *out = tmp1 + tmp2;
    }
}

int main(void)
{
    int var = 50;
    int result ;
#pragma css start
    fib (var-1, &result);
#pragma css finish
    printf ("fib(%d) = %d\n", var,  result );
    return 0;
}
```

Listing 11.1 StarSs example of parallel, recursive Fibonacci computation

descriptors, and most importantly the memory address space. In order not to create race conditions, multiple threads need to synchronize access to shared resources.

The StarSs programming model [5] allows to specify tasks and their interdependencies by annotating the sequential code using pragma statements (in C) and comments (Fortran). Code 25 shows an example written in C, with the StarSs statements being followed by the `css` pragmas.[2]

While compilers not supporting StarSs ignore the pragmas in Code 25, BSC's SmpSs compiler generates wrapper functions out of the specified tasks, with the StarSs runtime scheduling the tasks to threads calling these wrapper functions, which then invoke the underlying code generated by the host-compiler. The threads are beingcreated upon the entering of the `start` section. Currently, the C, C++ and Fortran77 and Fortran90 languages are supported – of course, due to it's thread-parallel nature, several unsafe practices may not be used in the taskified code, such as C's long-jumps, unprotected access to shared resources (mainly global memory), C++ exceptions, or in Fortran unprotected access to common blocks or passing temporary arrays. As with OpenMP the details of the underlying Thread-model (Posix Threads, Solaris Threads) [4] as well as the semantics of the Memory Model [1] are left to the compiler. Similar to OpenMP, the StarSs programming

[2]The acronym `css` is due to BSC first developing the SuperScalar compilers for the IBM Cell.

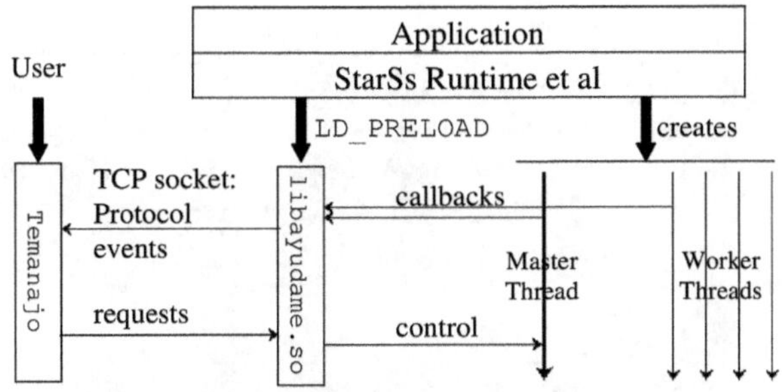

Fig. 11.1 Interaction between StarSs runtime, Ayudame and Temanejo

model allows incremental parallelization of the application, focusing first on large, independent tasks, leaving the source code's structure mostly intact – even more so, as soon as future compiler releases allow taskifying code sequences within functions.

The StarSs' runtime handles the dependency tracking using the parameter's address being passed to the task. As soon as all input dependencies are being met, the task may be scheduled to run. Compared to the classical OpenMP approach focusing on parallelising large loops, StarSs allows a much more dynamic execution, possibly making better use of resources. As will be shown below, the task-graph is a directed acyclic graph, which is dynamically generated at runtime. This graph may become very large and is executed non-deterministically.

However, this non-determinism may create problems when debugging the application. In this paper, we will revisit the Temanejo-debugger [2], developed at the High Performance Computing Center Stuttgart (HLRS) within the frame of the TEXT project.

11.2 Implementation

The graphical debugger Temanejo[3] is logically separated from the library that interacts with the runtime, the so-called Ayudame-library.[4] This separation allows to attach to SMPSs instances running on compute nodes, with limited capabilities, having the graphical display of tasks on the programmer's workstation. Ayudame provides multiple hooks for the runtime, e.g. to retrieve the number and names of tasks in the parallel section and API to steer the runtime. Figure 11.1 shows the interaction of the library Ayudame, here being `LD_PRELOAD`ed to the run-time, in order to satisfy the callbacks into the hooks. The Master Thread here stands for

[3]Spanish for 'I Handle You'.

[4]Spanish for 'Help me'.

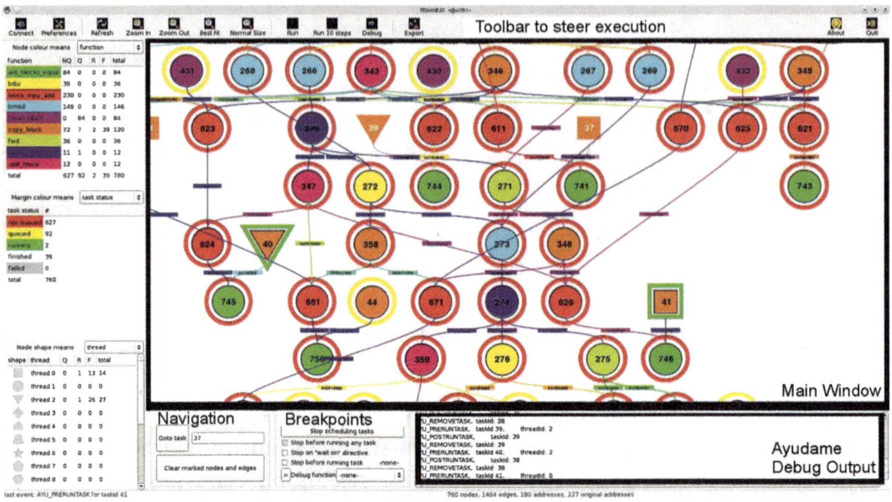

Fig. 11.2 Screenshot of the Temanejo GUI zoomed in, while debugging a LU-factorization example

SMPSs' runtime scheduling the tasks. The data exchanged between the Ayudame library and the Temanejo debugger is kept lean as a binary protocol, in principle allowing to attach and detach from the debugged process. The communication is currently done using TCP. If the user issues commands from the Temanejo GUI, it is being forward to the Ayudame library as a requests, which then relays it to the runtime. All the events on the run-time site are being collected and buffered to be sent in chunks via a separate thread decoupling the runtime and the GUI. On the GUI's side again data is received in a separate thread analyzing the event changes and updating the graphical representation.

11.3 Debugging Capabilities

Debugging implies seeing, understanding and intercepting the execution of an application. The Temanejo debugger allows programmers to visualize even large task graphs and efficiently interact with the run-time to analyze large applications.

Figure 11.2 presents a screenshot (edited for better printing) of Temanejo, here with full detail and heavily zoomed to enlarge the features of this graph.

Each task is visualized by a node in the main window of Temanejo. The user may zoom and drag through the main window and click on single nodes, aka the tasks, offering further functionality as described later. The node's color and shape by default encode the task's function and when the task is scheduled and running, the actual thread number it is running on. Moreover, the screenshot in Fig. 11.2 additionally shows the dependencies including the address of the dependency.

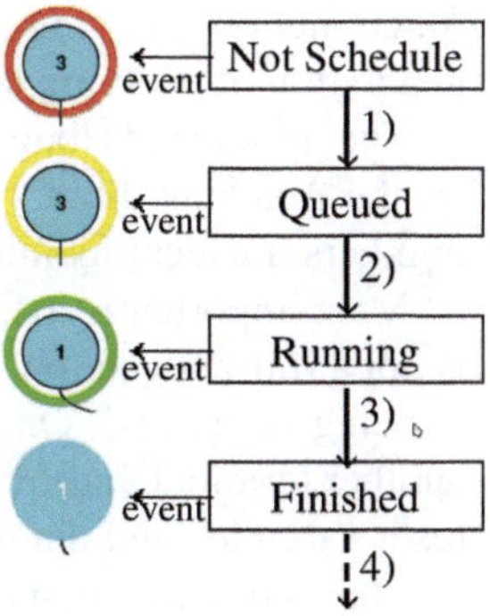

Fig. 11.3 Symbols representing the states within the life-time of a task, each change in state will generate an event in Ayudame

This nicely visualizes the applications capabilities in terms of parallel execution, e.g. narrow graphs represent situations in the execution where little opportunities for parallelism exist.

The node's border or margin encodes the tasks state Fig. 11.3 shows Temanejo's representation of the task state life-cycle. Tasks change their state upon the following conditions and emit the following event (numbering corresponding to Fig. 11.3):

1. If all input dependencies are fulfilled, the task may be enqueued.
2. If the task is being dequeued from the runable queue it is scheduled for execution to a thread.
3. If the task is finished running, another event is being generated to update the display. In reality it is being dropped from all queues.
4. At last, the tasks depending on the output of the finished tasks are being notified.

With the left-most pane of the display, the visualization attributes of the main window may be changed. Switching away from the default of the task's function, one may encode the executing thread (instead of it's shape), to distinctly visualize the likelihood of threads executing "close" tasks (the SMPSs runtime assumes dependent tasks benefit from data reusing the cache). Furthermore, one may set the node's color as it's execution time, showing specifically long-running tasks after the execution. Within large numbers of tasks the graph may get very bloated. Using the navigation controls, one may jump to a specific task and within the left-most pane change the node's color to encode the distance to the selected node allowing visual control of which subtree of the graph will remain dominant consumer of CPU time until the node in question will be executed.

In the toolbar, the application may be steered. By default, after attaching, the runtime is stopped from executing further tasks. The user may single-step through the application, receiving control over the SMPSs runtime after each schedule. Furthermore one may execute ten tasks at a time (or a setable number of steps) and regain control afterwards with a possibly drastically changed graph (due to new, added nodes, or lots of dependencies being cleared and many more tasks in runnable state). A nice feature to interact with the application developer is the export functionality, allowing single screenshots of the current graph and even movie generation of evolving task graphs.

The Ayudame Debug output in the lower right corner shows all the events emitted by the Ayudame library in human readable text, allowing easy searching for task numbers if executing multiple tasks at a time.

Most importantly a debugger will allow steering the application. Since Temanejo is not a full-fledged low-level debugger, the underlying system-debugger, here gdb, is being integrated. One may at any time attach to a task with the gdb, opening another terminal and from there use the usual techniques to query memory's state, reset variables and single-step through the binary. Temanejo helps in that regard, that the actual function and not the wrapper function is being attached to.

11.4 Related Work

Debugging multiple threads with it's many types of possible faults allows for a multitude of solutions. The standard tools required for debugging are traditional debuggers such as the GNU debugger (`gdb`); it allows switching between multiple threads or issuing specific commands using the `thread` keyword. However, to `gdb` the tasklets being created are functions of different name (including line number information).

Temanejo aims at a higher level visualizing and steering the task-graph's execution. Still, it relys on the underlying debugger, like `gdb` to attach to the generated functions to debug the actual source code.

An important class of faults are unprotected access to memory shared between threads. As of now, this class of bugs may not be found with our tool. However, using techniques similar to the valgrind- and pin-based tools developed in [3], accesses to global memory areas registered prior to starting the tasks could be tracked and checked for unordered access. There has been recent work by BSC to detect memory overrun and access errors of parameters passed to tasklets using the `valgrind`-tool.

11.5 Outlook

The tool has proven useful to programmers of StarSs-parallelized applications. We aim to extend the tool in other scenarios, such as debugging OpenMP tasking constructs, even though much of the visual capabilities for dependency tracking is lost. There are endeavours to abstract the Ayudame library in order to fit into a Pthread-based custom runtime system in order to visualize a large-scale numerical simulation. Moreover we hope to make good use of extended three-dimensional visualization, where one attribute leads to different panes in the third dimension, allowing spatial separation for better control of large graphs.

Acknowledgements We would like to thank Barcelona Supercomputing Center (BSC) and the partners in the TEXT project for their support in the development of their compiler suite.

References

1. Adve, S.V., Boehm, H.J.: Memory models: a case for rethinking parallel languages and hardware. Commun. ACM **53**(8), 90–101 (2010). doi: 10.1145/1787234.1787255
2. Brinkmann, S., Gracia, J., Niethammer, C., Keller, R.: Temanejo – a debugger for task based parallel programming models. In: Proceedings of ParCo'11, Gent, vol. abs/1112.4604, Gent, Belgium (2011)
3. Fan, S., Keller, R., Resch, M.: Advanced memory checking frameworks for MPI parallel applications in Open MPI. In: Tools for High Performance Computing. Springer, Berlin (2011). Submitted for publication
4. Northrup, C.J.: Programming with UNIX Threads. Wiley, New York (1996)
5. Perez, J.M., Badia, R.M., Labarta, J.: A dependency-aware task-based programming environment for multi-core architectures. IEEE Int'l Conference on Cluster Computing (Cluster 2008), Tsukuba, pp. 142–151 (2008)

Chapter 12
HiFlow^3: A Hardware-Aware Parallel Finite Element Package

H. Anzt, W. Augustin, M. Baumann, T. Gengenbach, T. Hahn, A. Helfrich-Schkarbanenko, V. Heuveline, E. Ketelaer, D. Lukarski, A. Nestler, S. Ritterbusch, S. Ronnas, M. Schick, M. Schmidtobreick, C. Subramanian, J.-P. Weiss, F. Wilhelm, and M. Wlotzka

Abstract The goal of this paper is to describe the hardware-aware parallel C++ finite element package HiFlow^3. HiFlow^3 aims at providing a powerful platform for simulating processes modelled by partial differential equations. Our vision is to solve boundary value problems in an appropriate way by coupling numerical simulations with modern software design and state-of-the-art hardware technologies. The main functionalities for mapping the mathematical model into parallel software are implemented in the three core modules Mesh, DoF/FEM and Linear Algebra (LA). Parallelism is realized on two levels. The modules provide efficient MPI-based distributed data structures to achieve performance on large HPC systems but also on stand-alone workstations. Additionally, the hardware-aware cross-platform approach in the LA module accelerates the solution process by exploiting the computing power from emerging technologies like multi-core CPUs and GPUs. In this context performance evaluation on different hardware-architectures will be demonstrated.

Keywords Parallel finite element software • High performance computing • Numerical simulation • Hardware-aware computing • GPGPU

12.1 Introduction

Scientific computing and numerical simulations are very important tasks in research, engineering and development. Mathematics combined with new software design dedicated to state-of-the-art hardware technologies result in high complexity but

H. Anzt · W. Augustin · M. Baumann · T. Gengenbach · T. Hahn · A. Helfrich-Schkarbanenko · V. Heuveline (✉) · E. Ketelaer · D. Lukarski · A. Nestler · S. Ritterbusch · S. Ronnas · M. Schick · M. Schmidtobreick · C. Subramanian · J.-P. Weiss · F. Wilhelm · M. Wlotzka
Engineering Mathematics and Computing Lab (EMCL), Karlsruhe Institute of Technology (KIT), Karlsruhe, Germany
e-mail: vincent.heuveline@kit.edu

H. Brunst et al. (eds.), *Tools for High Performance Computing 2011*,
DOI 10.1007/978-3-642-31476-6_12,

also break new grounds in the area of scientific computing. HiFlow^3 [4] is a parallel hardware-aware finite element software with the goal to provide powerful tools for the solution of complex problems arising in the area of medical engineering, meteorology and energy research. A team at the Engineering Mathematics and Computing Lab (EMCL) at the Karlsruhe Institute of Technology (KIT) is actively developing the software package. The goal of this paper is to present the concept of the software and the three core modules. Benchmark results for an advection-diffusion example comparing different platforms and preconditioners illustrate its usability and flexibility.

The paper is structured as follows. Chapter 2 outlines the design of HiFlow^3. The core modules are described in more details in Chaps. 3–5. Finally, benchmarks results for the advection-diffusion example are shown in Chap. 6.

12.2 Motivation, Concepts and Structure of HiFlow^3

12.2.1 Fields of Application

A typical application in the research area of medical engineering incorporating the full complexity of physical and mathematical modeling is the *United Airways* project [2]. Within an interdisciplinary framework its objective is the simulation of the full human respiratory tract including complex and time-dependent geometries with different length scales, turbulences, fluid-structure interaction (e.g. fine hairs and mucus), and effects due to temperature and moisture variations. Due to the complex interaction of the different physical effects it is a great challenge to define robust numerical methods giving accurate simulation results.

12.2.2 Flexibility

The conceptual goal of HiFlow^3 is to be a flexible multi-purpose software package. Its modular, transparent and documented structure tries to enable the user to set up a wide range of scenarios with reasonable effort. The core of HiFlow^3 is divided into three main modules: Mesh, DoF/FEM and LA; see Fig. 12.1. These three core modules are essential for the solution procedure which is based on the finite element method (FEM) for partial differential equations (PDEs). They are supplemented by other building blocks consisting of e.g. routines for numerical integration, matrix assembly, inclusion of boundary conditions, setting up nonlinear and linear solvers, providing data output for visualization of solutions and error estimators.

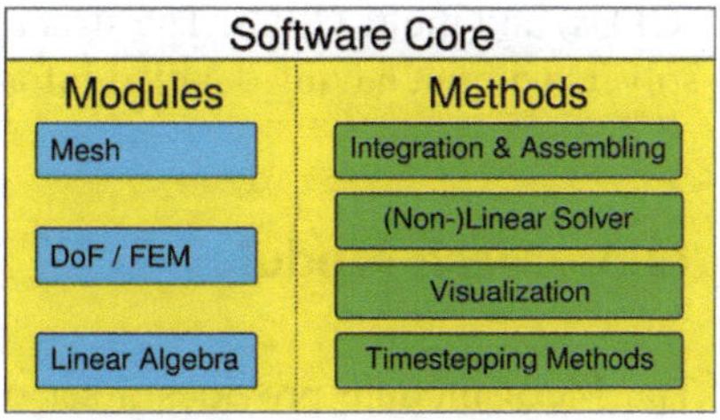

Fig. 12.1 Structure of the HiFlow^3 core divided into modules and methods

12.2.3 Performance, Parallelism, Emerging Technologies

The huge demand on fast and accurate simulation results for large-scale problems poses considerable challenges on the implementation on modern hardware. Supercomputers and emerging parallel hardware like GPUs offer impressive computing power in the range of Teraflop/s for desktop supercomputing up to Petaflop/s for cutting edge HPC machines. Therefore, an important goal associated with the design of HiFlow^3 is the full utilization of the available resources on heterogeneous platforms ranging from large HPC systems to a stand-alone workstation or a coprocessor-accelerated machine. To achieve this goal each core module is in itself parallel. An MPI-layer [12] realizes the communication between different nodes and processors. Furthermore, a hardware-aware computing concept, see Sect. 12.2.4, is implemented on the linear algebra level. This concept and all modules are laid out with respect to scalability in a generic sense. The design of HiFlow^3 aims at obtaining high standards with respect to the efficiency without sacrificing the flexibility of the software.

12.2.4 Hardware-Aware Computing

Hardware-aware computing is a multi-disciplinary approach to identify the best combination of applications, physical models, numerical schemes, parallel algorithms, and platform-specific implementations that gives the fastest and most accurate results on a particular platform [3]. Since hardware design and mathematical cognition give rise to different implementation concepts, hardware-aware computing also means to find a balance between the associated opponent guiding lines while keeping the best mathematical quality. All solutions need to be designed in a reliable, robust and future-proof context. The goal is not to design isolated solutions for particular configurations but to develop methodologies and concepts that preferably apply to a wide range of problem classes and architectures or that can be easily extended or adapted. The HiFlow^3 project realizes related concepts in the framework of the local multi-platform LAtoolbox [5, 6] within the LA module. Its use of specific implementations of basic routines like local matrix-vector and advanced preconditioning techniques is aimed at exploiting the available computing power of emerging technologies like multi-core CPUs, graphics processing units

(GPUs) and multi-GPUs. The structure provided by the LA module allows to build solvers without having detailed information on the underlying platform.

12.3 Mesh Module

The Mesh module provides a set of classes which can be used to represent the computational mesh in a finite element problem. In this section, the main ideas and features of the module are summarized. A more detailed description can be found in [13].

A mesh is a partitioning of a domain into a set of non-overlapping cells. Supported cell types are triangles and quadrilaterals in 2D, and tetrahedra and hexahedra in 3D. The description of the supported shapes are encapsulated in a class hierarchy, whereas the rest of the code is independent of the type and dimension of the cells.

The module can also represent lower-dimensional entities, such as edges and faces. The computation and storage of incidence relations between entities, which together make up the topology of the mesh, is performed in a way that closely follows the approach described in [11].

The geometry of the mesh is defined by assigning coordinates to each vertex, but the extension to more complete representations that could handle cells with curved faces is also possible.

The functionality provided through the Mesh module can be divided into three categories: queries, modifications and communication. The first category includes functions to iterate over the various entities and the topological incidence relations between them; as well as access to the geometrical information and other attributes associated with the entities such as material numbers. The modifying functionality includes creation, refinement and coarsening. A mesh can be created by specifying it entirely in code, or by reading a file. The AVS UCD [1] and VTK [15] formats are supported. For efficient I/O of large meshes, the module supports reading and writing of parallel VTK files. Refinement and coarsening does not have to be uniform, and there is functionality for dealing with the hanging nodes that arise as a result of local refinement.

For large-scale computations, the mesh might not fit into the memory of each process. For this reason, the HiFlow^3 Mesh module makes it possible to work with meshes that are distributed across a cluster of computers, as shown in Fig. 12.2.

Support for distributed meshes is provided by the functions in the communication layer. A distributed mesh is represented by a set of local mesh objects, one on each process. The communication functionality is external to these objects, which themselves are not aware that they are part of a larger, global mesh. This has the advantage that calls to query and modification functions are local to each process, and do not require communication, which facilitates their implementation and use.

The information about how the parts of the global mesh are connected is provided through a data structure that keeps track of which vertices are shared with other processes, and maps vertex identities between processes. With this structure on hand, it is also possible to identify shared entities of higher dimension.

Fig. 12.2 A mesh of a human nose distributed in 16 stripes

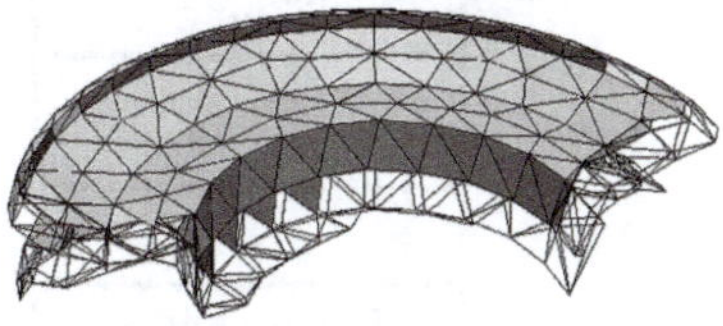

Fig. 12.3 A part of a mesh with the ghost layer

To facilitate the exchange of specific information required for finite element assembly, the Mesh module can also compute and communicate a layer of overlapping "ghost cells", which can be appended to each local mesh (see Fig. 12.3).

Different strategies can be used when deciding how to divide a mesh between the processes. Currently, the best option is to link via a provided interface to the popular graph partitioning library METIS [10].This enables the computation of a well-balanced partitioning of the mesh over the available processes.

12.4 DoF/FEM Module

In the context of FEM solutions are expressed in terms of linear combinations of some chosen shape functions defined on mesh cells. The degrees of freedom (DoF), which can easily count up to millions of unknowns, represent the number of parameters that define such a discrete function. To this end, local basis functions are defined on reference cells which are mapped into the physical space using transformations, see Fig. 12.4.

12.4.1 Submodules

The FEM submodule is dedicated to represent a finite element ansatz in a generic way to ensure extensibility at low memory costs, but still to enable high performance assembly. The basic concept lies in defining three major classes `FEType`,

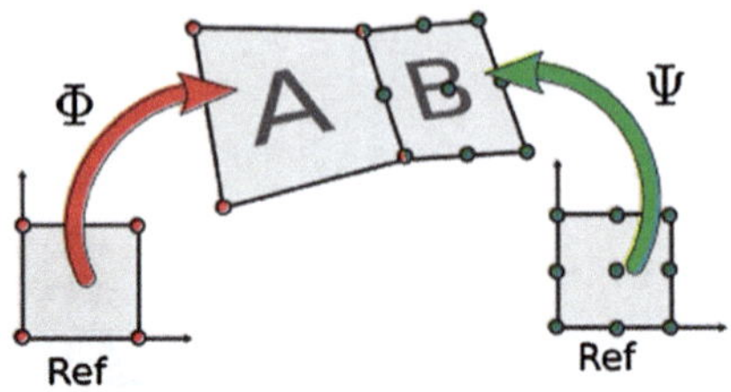

Fig. 12.4 Transformations Φ and Ψ map DoF points of biliinear and biquadratic finite element ansatz from reference cells to cells A and B of the mesh, respectively

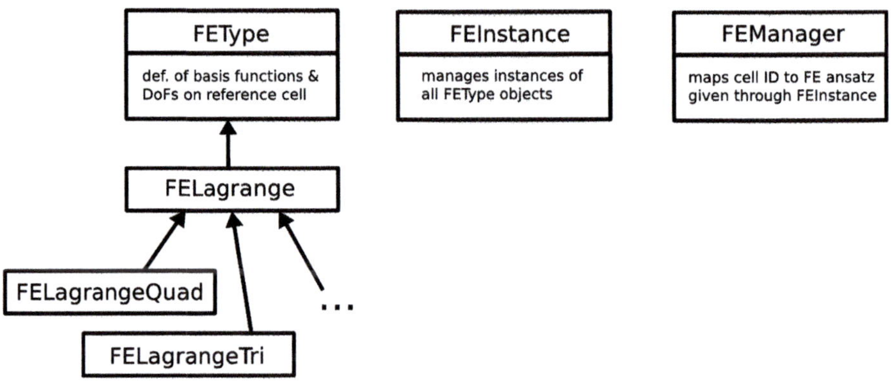

Fig. 12.5 Overview of classes for description of finite element functions

`FEManager` and `FEInstance`, see Fig. 12.5. `FEType` is an interface class that represents a specific continuous or discontinuous finite element ansatz on a defined reference cell by utilizing the design-pattern *singleton* stored within `FEInstance`. For establishing the mapping between a tuple of a mesh cell and a variable to their corresponding singleton, only *references* are stored. This mapping and all interfaces to out-of-module classes are managed by the `FEManager`.

Many practical relevant problems require a great number of DoFs (up to millions). These DoFs need to be numbered and interpolated in a unique way depending on arbitrary combinations of finite element ansatz on neighboring cells. One major task of the DoF submodule is to create a mapping between a local DoF Id and a global (mesh wide) DoF Id given by the local numbering strategy in the FEM submodule. The second major task is to interpolate those DoFs, which are restricted due to conditions provided by FEM and cannot be identified. The DoFs in any cell of the considered mesh are determined by a transformation [14] that maps the reference cell to the chosen physical cell and thereby defines the location of the DoF points (see Fig. 12.4).

12.4.2 Partitioning

In a domain decomposition setup, i.e. several processes are used for the solution of one single domain, each part of the domain (subdomain) is dedicated to one process using MPI. Again a unique DoF numeration of all DoFs in the global domain must be determined. For good scaling properties, a parallel and distributed handling of the DoFs is needed, i.e. each process manages the DoFs that are connected to the cells lying in its domain and the class `DofPartition` is used to create the correspondence with other subdomains from other processors via MPI communication. Each subdomain has information of the neighboring domains by the *ghost cells* which is a read-only copy of the neighboring cells (one layer neighborship of the processor's subdomain). To create a DoF numbering with (domain-) global Ids, each process determines in a first step a consecutive numbering of the DoFs within its subdomain, whereas also the ghost layer is treated as if it would belong to the subdomain. The antiquated information stored in this layer must be updated via communication. Hereby, a decision needs to be made, whether a DoF lying on the skeleton of the domain belongs to a subdomain or not, i.e. this DoF is lying on two subdomains, which are sharing it. The implemented procedure states, that the subdomain represented by a unique lower subdomain Id will own the DoF.

12.5 Linear Algebra Module

The LA module handles the basic linear algebra operations and offers linear solvers and preconditioners. It is implemented as a two-level library. The upper (or global) level is an MPI-layer which is responsible for the distribution of data among the nodes and performs cross-node computations, e.g. scalar products and sparse matrix-vector multiplications. The lower (or local) level takes care of the on-node routines offering an interface independent of the given platform.

The MPI-layer of the LA module takes care of communication and computations in the context of FEM. Given a partitioning of the underlying domain (handed over by the mesh module, see Sect. 12.3) the DoF module (see Sect. 12.4) distributes the DoFs according to the chosen finite element approach. This results in a row-wise distribution of the assembled matrices and vectors. Each local sub-matrix is divided into two blocks: a diagonal block representing all couplings and interactions within the subdomain, and an off-diagonal block representing the couplings across subdomain interfaces.

The highly optimized BLAS 1 and 2 routines are implemented in the local multi-platform linear algebra toolbox (lmpLAtoolbox) which is acting on each of the subdomains. By using unified interfaces as an abstraction of the hardware we provide easy-to-use access to the underlying heterogeneous platforms. In this respect the application developer can utilize the LA module without any knowledge

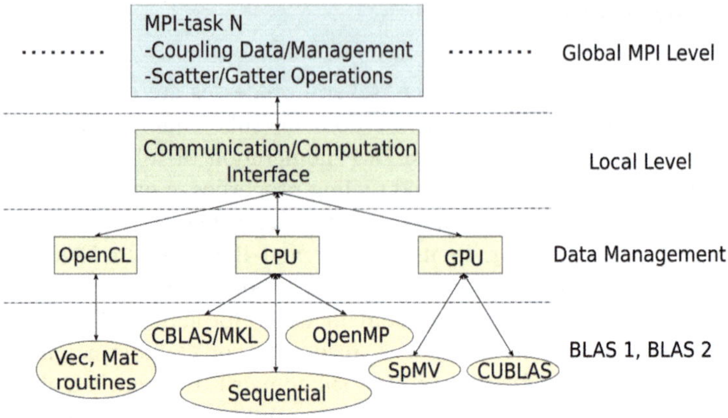

Fig. 12.6 Structure of the lmpLAtoolbox and LA module for distributed computation and node-level computation across several devices in a homogeneous and heterogeneous environment

of the hardware and the system configuration. The final decision on which specific platform the application will be executed can be taken at run time. Currently, the lmpLAtoolbox support sequential, OpenMP parallel and Intel MKL CPU versions, NVIDIA GPU version and OpenCL back-end which provides access to ATI and NVIDIA GPUs. Due to the flexible design of the library we can add a new back-end and still use the same solvers without modifying them.

The layered structure and organization of both the LA module and the lmpLA-toolbox is depicted in Fig. 12.6. It shows a high-level view of distributed communication and computation across nodes and the node-level computation across several devices in a heterogeneous environment. Details on software description and on the design, as well as performance benchmarks on several platforms can be found in [5, 6, 9].

Due to the modular setup and the consistent structure, the lmpLAtoolbox can be used as a standalone library independently of HiFlow^3. It offers a complete and unified interface for many hardware platforms. Hence, the LA module not only offers fined-grained parallelism but also flexible utilization and cross-platform portability.

12.5.1 Linear Solvers and Preconditioners

The LA module offers a high-level of abstraction by providing unified interfaces for basic matrix and vector routines. Platform-specific implementations are transparent to the user. Thus, linear and non-linear solvers can be implemented easily and generically without any information on the underlying hardware platform while keeping platform-adapted and tuned code. The default solvers in HiFlow^3 are preconditioned and non-preconditioned Krylov subspace solvers like – CG and

GMRES. For the GPU implementation of the CG solver we offload all of the vector and matrix operations on the device [5, 6], while for the GMRES solver we use hybrid-parallel approach where the basic vector and matrix routine are offloaded but the Krylov subspaces in still build on the CPU [9].

On the node level, we provide very fined-grained parallel preconditioners based on additive, multiplicative (incomplete LU-factorization) and approximate inverse techniques. In order to obtain high degree of parallelism we use multi-coloring decomposition of the matrix for the additive preconditioner and for the ILU(0). For the incomplete factorization with fill-ins we provide new technique for controlling the additional fill-in entries – power(q)-pattern method [8]. By using this algorithm we can execute in parallel the forward and backward sweeps of the LU matrix. An example for a Poisson problem solved with matrix-based multigrid method and with preconditioned CG solver based on HiFlow^3 are reported in [7].

12.6 Example: Advection-Diffussion Equation

The goal of this section is to show the simulation workflow in HiFlow^3 and benchmark results for different platforms and preconditioning techniques. As an example the well known advection-diffusion equation is chosen. We aim to find $u \in H_0^1(\Omega) := \{v \in H^1(\Omega) : v|_{\partial\Omega} = 0\}$ such that

Algorithm 3 Overall structure of the advection diffusion simulation

Read mesh, refine and distribute to all processors (Mesh).
Number the DoFs using a defined finite element space (DoF/FEM).
Assemble the system matrix and right-hand-side vector (LA).
Incorporate Dirichlet boundary conditions.
Prepare preconditioners.
Solve system by (preconditioned) linear solver.
Visualize solution.

$$\epsilon(\nabla u, \nabla \phi) + (\beta \cdot \nabla u, \phi) = (f, \phi) \quad \forall \phi \in H_0^1(\Omega), \tag{12.1}$$

where $(\cdot,\cdot)$ denotes the $L^2(\Omega)$ scalar product, $\Omega \in \mathbb{R}^d$ $(d = 2, 3)$ is a Lipschitz domain and $f \in L^2(\Omega)$ given. $\epsilon \in \mathbb{R}$ represents the diffusion coefficient and the vector $\beta \in \mathbb{R}^d$ the direction of transport and information flow. Numerical oscillations may occur for small diffusion coefficients, i.e. $\|\beta\|_2 \gg \epsilon$. In order to prevent this, ϵ and β are chosen depending on the characteristic mesh size h such that for the Peclet number Pe (see [9] and references therein) holds:

$$Pe = \frac{h\|\beta\|_2}{\epsilon} < 1.$$

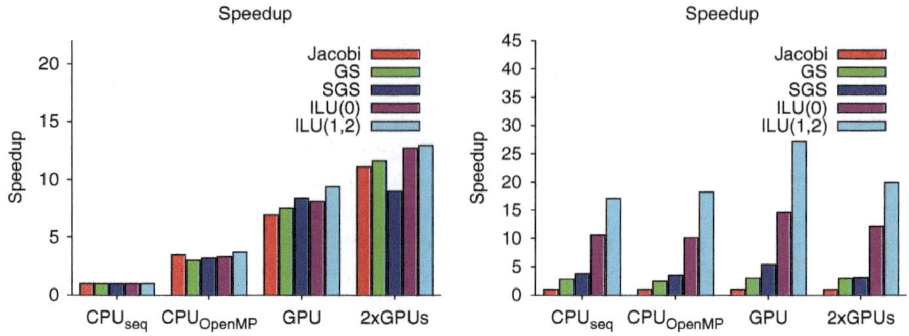

Fig. 12.7 Speedup comparing different platforms (eight cores CPU, 1× GPU, 2 × GPU) w.r.t. to single core CPU (*left*) and speedup of different preconditioners (Multi-colored (symmetric) Gauss-Seidel and ILU(0), and power(q)-pattern ILU(1, 2)) on the four different platforms w.r.t. Jacobi preconditioner (*right*)

For the finite element solution procedure the following steps have to be performed to map the weak formulation of advection diffusion example (12.1) to HiFlow^3, see Algorithm 3. In a preprocessing step the domain of interest is discretized by means of a finite element triangulation. The spatial grid is read in, adjusted by global refinement, partitioned using the external library METIS [10] and distributed via MPI. Then, a problem-adapted finite element space with appropriate ansatz functions is chosen. Once these two information are provided a unique global numbering of the DoFs is determined. In the case of distributed memory platforms the DoF-partitioner is used to create the global numbering throughout the whole computational domain by only using its local information and MPI communication across the nodes. In the next step the linear non-symmetric system matrix and the vector for the right hand side can be assembled by local integration over all elements within the triangulation. Data structures for matrices and vectors are provided by the LA module. An adequate linear solver for non-symmetric matrices is combined with different preconditioners to solve the linear system. Finally, output in either sequential or parallel format is provided for the visualization.

12.6.1 Numerical Results and Benchmarks

The numerical experiments on multicore-CPUs and multi-GPU configurations are performed on a dual-socket Intel Xeon (E5450) quad-core platform equipped with an NVIDIA Tesla S1070 system providing two GPUs per socket. The memory capacity of the host CPU system is 16 GB and 4 × 4 GB for the GPUs. We are only using two GPUs associated with a single socket. All simulations are performed

Table 12.1 Number of iterations and solving time of the preconditioned GMRES solver. With (*) we denote when the number of iterations exceeds 100,000 and in this case we stop the solver

		No precond	Jacobi	GS	SGS	ILU(0)	ILU(1, 2)
Sequential	# its	*	40,103	10,035	6,122	2,185	792
	time [s]	28179.3	11722.7	4132.7	3097.6	1105.8	686.1
OpenMP	#its	*	41,871	10,850	6,238	2,185	792
	time	7301.0	3364.1	1381.1	970.8	333.8	184.7
GPU	#its	84,800	44,106	10,642	6,050	2,243	792
	time [s]	3747.8	1982.2	550.7	369.9	136.1	73.2
2 GPUs	#its	87,503	41,618	12,176	10,024	2,571	1,031
	time [s]	2889.2	1057.5	356.5	344.0	87.1	53.2

in double precision. For the numerical experiments we consider an exact solution given in 2 dimensions by

$$u(x,y) = x\frac{e^{\beta_1 x} - e^{\beta_1}}{e^{\beta_1}}\sin(\pi y), \qquad (x,y) \in \Omega = (0,1)\times(0,1).$$

The unit square is discretized by quadrilaterals and globally refined such that $h = \frac{1}{2^9}$. Q_2 Lagrangian finite elements ansatz function are chosen, which leads to $1,025^2 = 1,050,625$ DoFs. The non-symmetric system matrix has $16\cdot 10^6$ non-zero entries. We use the iterative (preconditioned) GMRES method to solve the problem [10]. The stop criteria for the solver is set to a relative tolerance of 10^{-6}, which results for the chosen benchmarks settings in an absolute tolerance about 10^{-9}.

The number of iterations and the performance times for the sequential, OpenMP, one and two GPUs are shown of Table 12.1. Clearly, the problem is very ill conditioned and the system is hard to solve without preconditioning. When solving this system on different parallel architectures due to the parallel reduction and the different distribution of the floating-point errors, we obtain different number of iterations. This may be avoided by applying a 'good' preconditioner, leading to a well-conditioned system. The corresponding speedup factors for parallel execution and for different preconditioners are presented on Fig. 12.7. The speedup depends on the utilization of the available bandwidth of the platforms – with one core we can utilize one fourth of the bandwidth and therefore the OpenMP implementation is three to four times faster then the sequential version on the CPU. The GPU implementation however is about 9–13 times faster. The speedup profile of the preconditioners is the same for the sequential, OpenMP and GPU version – this is an important characteristic predominant with the high degree of parallelism of our preconditioners.

For the two GPUs case we apply the preconditioners in the block-Jacobi fashion with two blocks for each GPU. Doing so we decrease the coupling of the precondition matrix, thereby increasing the number of iterations. Even though with two GPUs we may utilize twice the bandwidth, this limits the speedup to less than two.

12.7 Conclusion

The creation of a multi-purpose finite element software package that is portable across a wide variety of platforms including emerging technologies like hybrid CPU and GPU platforms is a challenging and multi-facetted task. The modules Mesh, DoF/FEM and Linear Algebra complemented by auxiliary routines and interfaces provide a broad suite of building blocks for development of modern numerical solvers and application scenarios. The user is freed from a detailed knowledge of the hardware – one only has to familiarize with the provided interfaces and needs to customize the available modules in order to adapt HiFlow^3 to his domain-specific problem settings.

With hardware-aware solvers HiFlow^3 allows to compute the linear system of differential equations on different back-ends with single source code for all platforms. Besides the solvers, HiFlow^3 provides very fined-grain parallel preconditioners for multi- and many-core platforms and to our best knowledge is the first software package providing incomplete LU factorization preconditioners on the GPU devices.

Acknowledgements The *Shared Research Group 16-1* received financial support by the Concept for the Future of Karlsruhe Institute of Technology in the framework of the German Excellence Initiative and the industrial collaboration partner Hewlett-Packard. The *United Airways* project thanks the Städtisches Klinikum Karlsruhe for providing us with CT-data for the simulations of medical processes.

References

1. Advanced Visual Systems [AVS]: http://help.avs.com/Express/doc/help/reference/dvmac/UCD_Form.htm
2. Baron, L., Gengenbach, T., Henn, T., Heppt, W., Heuveline, V., Kratzke, J., Krause, M.J.: United airways: numerical simulation of the human respiratory system. http://www.united-airways.eu (2011)
3. Buchty, R., Heuveline, V., Karl, W., Weiss, J.P.: A Survey on Hardware-aware and Heterogeneous Computing on Multicore Processors and Accelerators. EMCL Preprint Series. KIT, Karlsruhe (2009)
4. Heuveline, V., et al.: HiFlow^3 – A Flexible and Hardware-Aware Parallel Finite Element Package. EMCL Preprint Series. KIT, Karlsruhe (2010)
5. Heuveline, V., Lukarski, D., Weiss, J.P.: Scalable multi-coloring preconditioning for multi-core CPUs and GPUs. In: UCHPC'10, Euro-Par 2010 Parallel Processing Workshops, vol. 6586, pp. 389–397. Springer/LNCS, Heidelberg (2010)
6. Heuveline, V., Subramanian, C., Lukarski, D., Weiss, J.P.: A multi-platform linear algebra toolbox for finite element solvers on heterogeneous clusters. In: PPAAC'10, IEEE Cluster 2010 Workshops. Heraklion, Crete, Greece (2010).
7. Heuveline, V., Lukarski, D., Trost, N., Weiss, J.P.: Parallel Smoothers for Matrix-Based Multigrid Methods on Unstructured Meshes Using Multicore CPUs and GPUs. EMCL Preprint Series. KIT, Karlsruhe (2011)

8. Heuveline, V., Lukarski, D., Weiss, J.P.: Enhanced Parallel ILU(p)-Based Preconditioners for Multi-core CPUs and GPUs – The Power(q)-Pattern Method. EMCL Preprint Series. KIT, Karlsruhe (2011)
9. Heuveline, V., Subramanian, C., Lukarski, D., Weiss, J.P.: Parallel preconditioning and modular finite element solvers on hybrid CPU-GPU systems. In: Proceedings of ParEng 2011, Paper 36. Civil-Comp Press, Stirlingshire (2011)
10. Karypis, G., Kumar, V.: A fast and highly quality multilevel scheme for partitioning irregular graphs. SIAM J. Sci. Comput. **20**(1), 359–392 (1999)
11. Logg, A.: Efficient representation of computational meshes. Int. J. Comput. Sci. Eng. **4**(4), 283–295 (2009)
12. MPI Forum: MPI: a message-passing interface standard. Version 2.2. Available at: http://www.mpi-forum.org (2009)
13. Ronnas, S., Gengenbach, T., Ketelaer, E., Heuveline, V.: Design and Implementation of Distributed Meshes in HiFlow3. In: Proceedings of CiHPC 2010, Schwetzingen. Proceedings of CiHPC 2011. (accepted)
14. Schieweck, F.: A General Transfer Operator for Arbitrary Finite Element Spaces. Magdeburg University, Magdeburg (2000)
15. VTK – The Visualization Toolkit: http://www.vtk.org/

Author Index

H. Brunst et al. (eds.), *Tools for High Performance Computing 2011*,
DOI 10.1007/978-3-642-31476-6,

GPSR Compliance

The European Union's (EU) General Product Safety Regulation (GPSR) is a set of rules that requires consumer products to be safe and our obligations to ensure this.

If you have any concerns about our products, you can contact us on ProductSafety@springernature.com

In case Publisher is established outside the EU, the EU authorized representative is:

Springer Nature Customer Service Center GmbH
Europaplatz 3
69115 Heidelberg, Germany

Batch number: 10584436

Printed by Printforce, the Netherlands